# Preface

This book began as the result of an initiative by the Structures Group of the Institution of Civil Engineers. It is intended as a design guide for engineers with a sound knowledge of structural design who are encountering the problem of seismic design for the first time.

Because national codes vary it has been difficult to produce a design approach which represents a common denominator for all codes. Some codes are referred to frequently and this reflects the author's personal experience to some extent, together with his view that these are competent codes. Many other excellent national and regional codes exist and their omission from this book should not be taken as a criticism. After all the current *World list of earthquake regulations* runs to 904 pages and an informed evaluation of them would be a formidable task.

Assistance in preparing the text and illustrations is gratefully acknowledged from many friends and colleagues. The library staff at the Earthquake Engineering Research Center, Berkeley, California, and at Imperial College engineering department library, London, have been of unfailing help. Colleagues at Consulting Engineers Partnership Ltd in Trinidad and Barbados have given advice and information, particularly Anthony Farrell, and their libraries have been a source of much useful reference work. Edmund Booth of Ove Arup and Partners, London, has acted as reviewer and his comments have been of great assistance.

In obtaining photographs much assistance has been given by members of the Earthquake Engineering Field Investigation Team, including Dr Colin Taylor and Edmund Booth. Jack Meehan of the California State Schools Department has kindly allowed me to reproduce photographs taken from some of his many field trips to major earthquake areas. Dr C. Browitt of the British Geological Survey in Edinburgh kindly arranged production of the world seismicity map.

# Contents

# Introduction

This book deals with earthquakes, which are natural disasters. In a letter to *The Times*, on 13 July 1984, the Archbishop of York wrote

'Disasters may indeed be messengers, in that they force us to think about our priorities. They drive us back to God. They remind us of mistakes and failures, and they call forth reserves of energy and commitment which might otherwise remain untapped. Disasters also remind us of the fragility of life and of our human achievements.'

Designing for earthquake resistance is difficult, not because the basic steps in the process are necessarily hard, but because the fundamental concept of earthquake resistance is different from design for other loadings, such as wind pressure or gravity loads. It is different in two important respects. Firstly it is a dynamic loading involving a number of cyclic reversals, so that the behaviour of the structure involves an understanding of structural dynamics. Secondly, normal design practice accepts that, in response to a major earthquake, a building structure may suffer major damage (but should not collapse), whereas for wind and gravity loads even minor damage is not acceptable.

Earthquake-resistant design is not widely taught. For the practising engineer it is a difficult subject to come to grips with, not because there is a shortage of information, but because there is a surfeit. It is a subject where it is possible to drown in information and to starve for knowledge. Professor G. Housner, in an address to the participants at the Eighth World Conference on Earthquake Engineering in 1984, suggested that, if the current logarithmic increase in the number of papers presented at the four-yearly World Conferences continued, by the 19th it would take four years to present the papers.

The author himself has struggled over many years to develop a sound approach to the design of structures in earthquake zones. This book is intended to guide others not only in the basic procedures of design but also to point out sources of specialised information on the subject when it is beyond the scope of this work.

There are many design codes available around the world, and it is not practicable to distil a common approach from them. In general it is assumed that the design engineer or architect reading this work will be following an appropriate code. Guidance on the appropriate features that are common to most codes is given in Chapter 6, but where specific recommendations for structural calculation are given they are drawn from American, New Zealand or, in some cases, European, practice. The reasons for this are twofold, the first being that they are widely accepted, and the second that they are among the most up to date at the time of writing. Many other good national codes exist. Codes should always be for the guidance of wise men, and compliance with a fixed set of rules does not guarantee adequate performance in a strong earthquake. Codes are all too commonly treated as maximum requirements, whereas they should be minimum requirements.

Earthquake engineering has to a large extent slipped out of the hands of the practical designer, and into the hands of the specialist, who usually employs a suite of computer programs to provide great quantities of unnecessarily precise information on such subjects as the ground motion spectrum or the dynamic response of the building to some long past earthquake which can only bear the vaguest resemblance to any ground motion to which the building could be subjected. In the author's view the principal ingredients in an earthquake-resistant design can be categorised as follows.

*Essential*
- (a) a sound structural concept
- (b) an understanding of the way in which the structure will behave when primary structural elements have yielded
- (c) an approximate idea of the peak ground acceleration likely to be experienced, and the predominant frequency
- (d) the application of engineering common sense to the fact that the building may be violently shaken
- (e) good detailing
- (f) good quality construction and inspection.

*Useful*
- (a) detailed elastic analysis of the structure
- (b) dynamic analysis of simple models
- (c) a soil–structure interaction study when justified by the soil and structure properties
- (d) estimates of the ground motion spectrum.

The designer is in the end the person who puts all the theory into steel and concrete, and who bears the responsibility for it.

This book assumes a competent knowledge of structural design by the reader. It is intended as a guide to the normal processes of design, and to provide directions for further study when the structural problem is out of the ordinary.

# Notation

$a_c$    attachment amplification

$a_{max}$    peak value of time-based record

$A_{ch}$    column section area measured to the outside of the transverse reinforcement

$A_g$    gross column section area

$A_j$    cross-sectional area of a joint

$A_p$    Fourier coefficient

$A_{sh}$    cross-sectional area of transverse reinforcement at spacing $s$

$A_v$    peak ground acceleration as a fraction of $g$

$b_w$    web thickness of a concrete section

$b^*$    shear hinge length

$c$    velocity of wave propagation in soil

$c_u$    undrained shear strength of soil

$[C]$    damping matrix

$C_c$    lateral force coefficient for a secondary element

$C_{mx}$    buckling analysis coefficient

$C_{my}$    buckling analysis coefficient

$C_p$    lateral force coefficient for a secondary element

$C_t$    response factor for building period calculations

$d$    pile thickness; effective depth to main reinforcement

$D$    building depth

$D_b$    external diameter of a chimney base (m)

$D_s$    shear wall depth

$D_t$    external diameter of a chimney top (m)

$e$    clear length of an EBF link beam; void ratio

$E$    Young's modulus

$f$    depth to the maximum bending moment; frequency

$f_c'$    specified concrete cylinder compressive strength

$f_{cb}'$    compressive strength of a concrete masonry unit

$f_e$    equivalent natural frequency

$f_g'$    compressive strength of concrete masonry grout

$f_i$    lateral force at level $i$

$f_m'$    prism strength of masonry

$f_y$    steel yield stress

$f_y^*$    factored steel yield strength allowing for overstrength

$f_{yh}$    yield strength of stirrups

$F$    force at the pile tip; friction force on the pipe per unit length

$F_p$    lateral force on a secondary element

$g$    acceleration due to gravity

$G$    shear modulus

| | |
|---|---|
| $G_0$ | scalar for ground motion |
| $G(\omega)$ | power spectrum |
| $h_i$ | height at level $i$ |
| $h_n$ | building height ($n$ storeys) |
| $h_x$ | height above the base to level $x$ |
| $h''$ | concrete core dimension to the outside of transverse reinforcement |
| $H$ | depth of soil |
| $H_u$ | ultimate lateral load |
| $H_x(\omega)$ | frequency response function |
| $H_0$ | height of the fixed fluid mass |
| $H_1$ | height of the vibrating fluid mass |
| $I$ | second moment of area |
| $I_x, I_y$ | second moment of area about $x, y$ axes |
| $k$ | stiffness |
| $k_e$ | equivalent stiffness |
| $k_s$ | coefficient of subgrade reaction |
| $K$ | structural type factor |
| $[K]$ | stiffness matrix |
| $[K^*]$ | modified stiffness matrix |
| $K_p$ | Rankine lateral pressure coefficient |
| $l_x, l_y$ | effective column lengths |
| $L$ | tank length; pile length |
| $m_b$ | body wave magnitude |
| $M$ | bending moment |
| $[M]$ | mass matrix |
| $M_p$ | plastic moment of resistance |
| $M_s$ | surface wave magnitude |
| $M_u$ | ultimate bending moment |
| $M_w$ | moment magnitude |
| $M_L$ | local magnitude |
| $M_0$ | fixed fluid mass |
| $M_{0x}$ | maximum column moment capacity, zero load |
| $M_1$ | vibrating fluid mass |
| $N$ | number of storeys; number of cycles |
| $N_u$ | factored axial load on a shear wall |
| $p_{sh}$ | lateral dynamic soil pressure |
| $p(t)$ | force at time $t$ |
| $p_u$ | ultimate soil resistance of cohesionless soil |
| $P$ | axial load on a column; performance criterion factor |
| $P_{ac}$ | allowable column load in compression |
| $P_e$ | design axial load |
| $P_j(t)$ | force at time $t$, normal coordinates |
| $P_t^*$ | modified load vector at time $t$ |
| $P_y$ | squash load on a column, $Af_y$; probability of occurrence in $y$ years |
| $r_x, r_y$ | radii of gyration, $x$ and $y$ axes |
| $R$ | return period; tank radius |
| $s_h$ | stirrup spacing |
| $S$ | soil factor |
| $S_a$ | spectral acceleration |

| | |
|---|---|
| $S_d$ | spectral displacement |
| $S_v$ | spectral velocity |
| $t$ | time; web thickness |
| $T$ | period of vibration (s); flange thickness |
| $T_c$ | fundamental period of a component |
| $v_p$ | compression wave velocity |
| $v_s$ | shear wave velocity |
| $V$ | shear; base shear |
| $W$ | mass |
| $W_p$ | mass of a secondary element |
| $x$ | displacement |
| $x_g$ | ground displacement |
| $x_0$ | displacement due to force $p_0$ |
| $Y_n$ | displacement, normal coordinates |
| $Z$ | seismic zoning factor |
| $\alpha$ | integration constant; constant; ratio of gross area to net area |
| $\beta$ | frequency ratio; ratio of column end moments |
| $\gamma$ | angle; soil strain; soil density; modal participation factor |
| $\delta$ | integration constant |
| $\Delta t$ | time increment |
| $\varepsilon$ | strain |
| $\theta$ | angle |
| $\mu$ | ductility factor |
| $\nu$ | Poisson's ratio |
| $\xi$ | fraction of critical damping |
| $\xi_e$ | equivalent fraction of critical damping |
| $\xi_g$ | fraction of critical damping, ground |
| $\rho$ | ratio of tension reinforcement; soil density |
| $\rho'$ | ratio of compression reinforcement |
| $\rho_s$ | steel volumetric ratio for stirrups |
| $\rho_w$ | ratio of reinforcement based on the web area |
| $\bar{\sigma}_m$ | mean principal effective stress |
| $\bar{\sigma}_v'$ | effective vertical overburden pressure |
| $\sigma_x$ | root mean square response of $x$ |
| $\sigma_y$ | yield stress in pure tension |
| $\tau$ | dummy time variable |
| $\tau_y$ | yield stress in pure shear |
| $\phi_n$ | $n$th mode shape |
| $\phi_0$ | overstrength factor in capacity design |
| $\Phi$ | modal matrix |
| $\omega$ | natural frequency; dynamic magnification factor in capacity design |
| $\bar{\omega}$ | forcing frequency |
| $\omega_d$ | natural frequency of the damped structure |
| $\omega_g$ | natural frequency of the ground |
| $\omega_i$ | natural frequency of the $i$th mode |
| $\omega_0$ | structure natural frequency |
| [ ] | square matrix |
| { } | vector |
| { }$^T$ | vector transposed |

# Chapter 1

# The lessons from earthquake damage

'The bookful blockhead, ignorantly read,
With loads of learned lumber in his head.'
*An essay on criticism*, Alexander Pope

## 1.1.  Damage studies

The study of earthquake damage was the original source of design criteria for earthquake-resistant structures. For example, following the 1906 San Francisco earthquake, engineers observed that buildings designed to withstand a wind force of 30 $lbf/ft^2$ performed well. That simple observation embodied a great deal of common sense, including the concept of using a static lateral force to reproduce the effect of an earthquake.

The reason for the quotation at the start of this chapter is to emphasise the author's view that earthquake engineering is not to be learned from books only. Engineers generally have some experience of their structures being loaded to something approaching their design load, and errors in calculation or implementation will show up in the form of major or minor defects. In this way there is some feedback from experience and some encouragement to use this experience. For earthquake design this is rarely the case, so that the only source of experience for an engineer is either the study of damage reports or, even better, in carrying out a damage survey himself. To take the subject of earthquake-resistant design out of the realms of a book-taught subject into the realms of thoughtful engineering it is essential that as much practical knowledge as possible is included. The designer needs to *feel* what may happen to one of his structures as well as to know a set of design rules.

Engineers are most accustomed to static loads. One of the most important lessons learned from damage surveys is the difference in failure patterns between static loads applied in a single direction and those due to cyclic loading. There are important differences in the way that crack patterns develop between the two.

Aftershocks, generally of much smaller magnitude than the main shock which they follow, play no explicit part in the design process. Nevertheless they play a significant part in the immediate post-earthquake rescue and survival operation, and the further damage that they do to already damaged buildings is greater than their magnitude would otherwise suggest. For a building to be seriously damaged requires a considerable amount of energy, and when this is used as a criterion, e.g. in the assessment of existing buildings, the additional input of energy from aftershocks should not be forgotten.

An important aspect of post-earthquake study is the realisation of the important role that the quality of construction plays. Earthquakes are no respecters of theories, calculations or divisions of responsibility. Many instances of poor quality construction are invariably exposed in damaged buildings. Badly placed reinforcement, poorly compacted concrete and incomplete grouting of masonry are some of the commonest examples.

The immediate human response to earthquakes is not in general regarded as a design criterion. Nevertheless every earthquake shows up numerous examples of lives at risk from minor faults in construction—falling masonry or cladding, ceiling tiles dislodged, window frames separating from the walls and toppling inwards or outwards, and escape paths blocked by jammed doors and fallen masonry.

In the longer term human response follows the pattern shown in Table 1.1, and while this might be seen as light hearted or cynical there is no doubt that, as the time of the last disaster recedes, it becomes increasingly difficult to convince owners, officials and pro-

Table 1.1.  *Long term human response to earthquakes*

| Stage | Time | Event | Reaction | |
|-------|------|-------|----------|--|
| | | | Positive | Negative |
| 1 | 0–1 minute | Major earthquake | | Panic |
| 2 | 1 minute–1 week | Aftershocks | Rescue and survival | Fear |
| 3 | 1 week–1 month | Diminishing aftershocks | Short term repairs | Allocation of blame—builders, designers, officials etc. |
| 4 | 1 month–1 year | | Long term repairs Action for higher standards | |
| 5 | 1 year–10 years | | | Diminishing interest |
| 6 | 10 years to the next time | | | Reluctance to meet costs of seismic provisions, research etc. Increasing non-compliance with regulations |
| 7 | The next time | Major earthquake | Repeat stages 1–7 | |

*Fig. 1.1. Local soil failure in reclaimed land at St Johns, Antigua, in 1974*

fessionals of the need for earthquake-resistant measures. The task of the building design team is not always neatly prescribed by sets of regulations, and the achievement of high technical standards requires a clear understanding of the problem and mutual support in presenting it to owners and officials.

## 1.2. Ground behaviour

The effects of violent shaking on the ground are temporarily to increase lateral and vertical forces, to disturb the intergranular stability of non-cohesive soils and to impose strains directly on surface material where the fault plane reaches the surface.

The results of a transient increase in lateral and vertical forces means that any soil structures that are capable of movement are at risk. The resulting types of damage are landslips and avalanches, and experience of the 1970 earthquake in Peru and the 1964 earthquake in Anchorage, Alaska, show that these may be on a massive scale. One village, Yungay, in Peru was destroyed almost entirely with the loss of 18 000 lives by a debris flow involving tens of millions of tons of rock and ice.

The disturbance of the granular structure of soils by shaking leads to consolidation of both dry and saturated material, due to the closer packing of grains. For saturated sands the pore pressure may be increased by shaking to the point where it exceeds the confining soil pressure, resulting in temporary liquefaction. This is an important effect as it can lead to massive foundation failure in bearing and piled foundations, the collapse of slopes, embankments and dams, and to the phenomenon of 'boiling' where liquefied sand flows upwards in surface pockets. It is also possible for some unstable soils to heave.

Shear movements in the ground may be at the surface or entirely below it. Where the earthquake fault reaches the surface permanent movements of considerable magnitude may occur, in metres rather

3

than centimetres (Fig. 1.1). Surface shear movements may also take place as a result of other soil displacement—landslips or consolidation for example. Subsurface shear failures can occur in weaker strata, leading to damage of embedded or buried structures. Subsurface shear failures can also reduce the transmission of ground motion to the surface, effectively putting an upper bound on the surface motion.

In considering the more spectacular permanent ground displacements that can result from ground shaking, it should not be forgotten that elastic displacements also occur and are critical in the design of piles, underground pipelines and culvert-type structures. Failures in underground piping and ductwork are common and have important implications in the post-earthquake emergency services.

### 1.3. Structural collapse

Figures 1.2–1.7 show some of the many ways in which structural collapse can occur. Collapse can initiate at any level and may be due to lateral or torsional displacement, local failure of supporting members, excessive foundation movement and occasionally the impact of another structure.

*Fig. 1.2. Failure of a multi-storey reinforced concrete structure in Mexico, 1985 (photograph by courtesy of M. Winney)*

*Fig. 1.3. Total collapse of a multi-storey reinforced concrete structure in Kalamata, Greece, 1986 (photograph by courtesy of A. Greeman)*

*Fig. 1.4. Intermediate collapse of a reinforced concrete framed structure in Mexico, 1985 (photograph by courtesy of M. Winney)*

5

*Fig. 1.5. Upper storey collapse of a multi-storey reinforced concrete structure in Mexico, 1985 (photograph by courtesy of M. Winney)*

*Fig. 1.6. Intermediate storey failure of a multi-storey structure in Mexico, 1985, probably caused, or aggravated, by buffeting against the adjacent building (photograph by courtesy of E. Booth, Earthquake Engineering Field Investigation Team)*

*Fig. 1.7. Partial collapse of a reinforced concrete apartment building in Mexico, 1985: failure of the end frame has precipitated local collapse (photograph by courtesy of C. Taylor, Earthquake Engineering Field Investigation Team)*

An important category of building failure is the case where it is so badly damaged that it has to be demolished, but does not collapse. For the owner and the insurance company the costs are similar whether the building collapses or is demolished. For the occupants it is the difference between life and death.

## 1.4. Important categories of damage

Where two buildings are close, or where there is a movement joint, the two sides are likely to pound against each other during an earthquake. Major structural damage can result from this, particularly where the floor levels differ. The cause lies in the closeness of the two structures and in the flexibility of the buildings, both of which are under the control of the designer.

Appendages to buildings—masonry parapets, penthouses, roof tanks, cladding and cantilevers—tend to behave badly. The reasons for this are twofold. Firstly many of them are designed without any ductility, and secondly the effects of dynamic amplification by the building to which they are attached may greatly increase the forces applied to them.

The contents of buildings often suffer major damage even when the building itself is relatively unharmed. This effect is greater for more flexible buildings and represents an additional reason for the designer to exercise close control over displacements. In many modern buildings the contents are of greater value and importance than the building itself. The costs of preventing damage are often

7

trivial—steel angle ties to the tops of racks, floor bolts to shelving for example.

Modern buildings are often assembled from many separate components. Older buildings also commonly have timber floors with joists that are poorly tied to the supporting walls. Any lack of tying together in a building is quickly exposed by earthquake shaking. The random nature of earthquake ground motion inevitably leads to differential movement between separate components, and in the absence of structural continuity differential movement will occur.

Anchorages of components into masonry or concrete by cast-in or expanding head bolts are almost invariably brittle in shear and tension, and thus unable to accommodate any movement. Accordingly failures are commonplace, aggravated when the masonry or concrete into which the anchorage is placed is damaged.

## 1.5. Framed structures

Framed structures will generally be engineered structures that are competent to deal with gravity and wind loads. In the familiar processes of design attention is commonly given to stresses rather than displacement. The secondary effects of displacement are forgotten. Earthquake damage frequently draws the attention back both to the direct effects of large displacements, such as the pounding at joints and damage to non-structural components and contents, and to the resulting secondary stresses. Buildings with shear walls or braced frames, as long as they maintain their integrity, compare favourably in performance with more flexible framed structures as far as damage to contents and non-structural items is concerned.

Particular points commonly brought out for framed structures in general include the following.

(a) Corner columns often behave badly in comparison with other exterior and interior columns. This suggests that the effects of earthquake forces in orthogonal directions are not adequately dealt with in design.

(b) Complete failure in members detailed for ductility is rare. Where members with low ductility have failed it is clear that deterioration is swift. This is particularly marked in reinforced concrete members.

(c) The maximum practicable redundancy is shown to be desirable. The failure mechanism should involve as many members as possible.

## 1.6. Non-framed buildings

Non-framed buildings will include a proportion of non-engineered buildings in which a calculation for resistance to gravity loads and wind is implicit rather than actual. Construction will be in various materials such as masonry, precast concrete, timber and traditional construction like adobe.

Many failures occur in horizontal torsion, especially in low-rise garage-type structures where the structure consists of a box with one side omitted. This places the centres of mass and resistance almost as far apart as is possible with a resulting high torsion.

Failures also occur due to a lack of tying the building together. This is particularly noticeable in structures incorporating timber diaphragms or floors. Frequently tying is limited to nails where it is dependent on the dowel action acting across the grain, or to nails in tension.

## 1.7. Reinforced concrete

Typical damage to elements subjected to bending, with or without direct force, includes (Figs 1.8–1.10)

(a) cracking in the tension zone
(b) diagonal cracking in the core
(c) loss of concrete cover
(d) the concrete core breaking into lumps by reversing diagonal cracking
(e) stirrups bursting outwards
(f) buckling of the main reinforcement.

In addition the following forms of failure occur

(a) bond failure, particularly in zones where there are high cyclic stresses in the concrete
(b) direct shear failure of short elements, or those constrained so that only a short length is effectively free

Fig. 1.8. Shear failure in a reinforced concrete column in Mexico, 1985: short lengths of column are liable to this brittle type of failure, particularly where the ends are constrained by heavy beams (photograph by courtesy of E. Booth, Earthquake Engineering Field Investigation Team)

*Fig. 1.9. Shear failure of a lightly reinforced concrete column and the adjacent masonry, opposite a window opening in St Johns, Antigua, 1974: the constraining effect of masonry abutting the column makes it liable to fail in shear before it fails in bending; at the weak point opposite the window opening the earthquake makes no distinction between concrete and masonry*

*Fig. 1.10. Reinforced concrete column in the ultimate stage of failure, after the San Fernando earthquake, 1971, at Olive View Hospital: gross displacement has occurred (photograph by courtesy of J. Meehan)*

(c) shear cracking in the beam–column intersection zone (panel zone)

(d) tearing of slabs at discontinuities, and junctions with stiff vertical elements

(e) diagonal cracking in shear walls, particularly concentrated around openings.

The causes and remedies for these various forms of structural distress are discussed in detail in Chapter 7.

## 1.8. Structural steelwork

Structural steel shows the following types of damage from earthquakes

(a) brittle failure of bolts in shear or tension

(b) brittle failure of welds, particularly fillet welds, in shear or tension

(c) member buckling, including torsional buckling

(d) local web and flange buckling

(e) uplift of braced frames

(f) local failure of connection elements such as cleats and T's

(g) bolt slip

(h) high deflections in unbraced frames

(i) failure of connections between steel members and other building elements, such as floors.

A detailed discussion of steelwork is given in Chapter 8.

## 1.9. Masonry

Failure of unreinforced masonry is so common that it is almost taken for granted and forgotten (Figs 1.11–1.14). Many earthquake

*Fig. 1.11. Typical X cracking in masonry panels in an Anchorage school, 1964 (photograph by courtesy of J. Meehan)*

11

*Fig. 1.12. Unreinforced masonry failed in Antigua, 1974, despite relatively low peak ground accelerations, probably less than 0·1g at this site: the potential for secondary damage in substations such as this is high*

*Fig. 1.13. Masonry failure in the San Fernando earthquake, 1971: the absence of horizontal reinforcement permits the formation of vertical cracks; the partial failure leaves a condition that was not anticipated by the designer with the masonry suspended from the eaves rafter (photograph by courtesy of J. Meehan)*

*Fig. 1.14. The separation of applied finishes such as this rendering to a church lintel in Antigua in 1974 gives rise to a hazard to occupants*

codes ban the use of unreinforced masonry altogether. However, economic reasons still ensure that it is widely used both for low-rise structural walls and as infill to framed structures.

Failures of both reinforced and unreinforced masonry in plane are common. In plane masonry is very stiff, so that the forces transmitted by ground shaking are high, and brittle so that failure is accompanied by a marked reduction in strength and stiffness. Damage normally comprises either collapse or diagonal cracking in both directions ('X' cracking) (Fig 1.11). Cracks will often be concentrated around openings. Cracking will frequently follow the mortar joints.

The effect of reinforcement on the in-plane damage is to reduce the amount of cracking and to reduce substantially the onset of failure.

Out of plane, free-standing masonry or masonry that has separated from any adjacent structure is liable to toppling failure. Toppling is much less likely if some mechanical connection exists at the sides and head of the wall. Reinforcement continued into a surrounding frame is most effective in avoiding complete collapse, acting as a basket to the masonry even when it is severely damaged.

Damage to masonry is often concentrated around openings.

The practice with unreinforced masonry in some places is to include lightly reinforced tie-beams and columns within panels. These undoubtedly reduce the amount of damage substantially.

## 1.10. Timber

Timber, like steel, is in itself an excellent material for earthquake resistance. Like steel it is in the connections where the weaknesses lie. Failures frequently occur in nailed joints where nails are in tension parallel to the grain or where edge distances are too small. Failures also occur in ledgers where nailed connections induce bending across the grain where timber is extremely weak.

13

Inadequate tying together of buildings occurs where there are timber floors. Unless positive connections are made at floor–wall junctions the walls can function as though there were no tie at all, and fail in out-of-plane bending. Timber floors do not function as rigid diaphragms and do not transmit lateral forces between stiff vertical elements as effectively as, for example, in situ reinforced concrete slabs.

## 1.11. Foundations

Failures of building foundations are not unusual but are nearly always caused by failure of the supporting soil (Figs 1.15–1.17). Overturning failures due to uplift occur rarely, far less often than calculations suggest. This is probably due to the effective reduction in stiffness that accompanies uplift, which correspondingly reduces the force exerted by the ground acceleration. There can be no doubt that substantial tension can develop at foundation level from a consideration of some lower columns in Caracas, following the 1967 earthquake. An examination of failed columns showed that they had failed in tension.

Following the San Fernando earthquake in 1971, excavations of piled foundations showed failures of piles at the junction with the base of the pile cap. Previously it had been considered that piles performed well. These piles were precast prestressed members with no specific provision for ductility. From this there is a good case for providing ductility in the upper portions of piles.

*Fig. 1.15. Although this structure had piled foundations, extensive weakening of the subsoil by the 1985 Mexico earthquake caused them to fail, producing a rotational failure (photograph by courtesy of M. Winney)*

*Fig. 1.16. This heavy structure has sunk by approximately one storey due to ground failure, in Mexico, 1985 (photograph by courtesy of M. Winney)*

*Fig. 1.17. Heavy settlement and adjacent street damage in the Mexico earthquake of 1985 (photograph by courtesy of E. Booth, Earthquake Engineering Field Investigation Team)*

15

## 1.12.  Non-structural elements

At any level on a multi-storey building the ground motion will be modified by the motion of the building itself. Generally the effect is to concentrate the frequency of response around a band close to the natural frequency of the building and to amplify the peak acceleration roughly in proportion to the height, reaching an amplification of perhaps two or three at roof level. For any contents which are either very stiff or which have a natural frequency of their own close to that of the building, this means that they are subjected to greater forces than they would be if mounted at ground level.

Experience shows that non-structural items which are suspended such as ceiling systems and light fittings perform badly. Appendages such as parapets also suffer high levels of damage, especially where they function as single degree of freedom inverted pendulums. Damage also increases on multi-storey structures towards the roof and roof tanks and penthouses are also subjected to high forces.

Other elements are damaged by drift, or interstorey displacement. Windows and cladding elements are frequently connected rigidly to more than one level and, if there is no ductile provision for movement in the connections, they will fail.

Examples of such damage are shown in Figs 1.18–1.21.

## 1.13.  Mechanical and electrical systems

Machinery mounted on antivibration springs performs badly in earthquake shaking (Fig. 1.22). Displacements are large and exceed

*Fig. 1.18. Although a building structure may sustain relatively light damage, the effect of displacement and acceleration forces on glazing may be substantial, as in this structure damaged by the Mexican earthquake of 1985 (photograph by courtesy of M. Winney)*

*Fig. 1.19. This masonry building in Kalamata, Greece, 1986, has remained essentially intact but has shed a considerable amount of external appendages such as canopies, cornices and parapets: it provides a warning against running out of a building during an earthquake (photograph by courtesy of A. Greeman)*

*Fig. 1.20. This roof tank failed during the Mexico earthquake, 1985: appendages like this suffer from the amplified response of the building and need to be designed with special care (photograph by courtesy of E. Booth, Earthquake Engineering Field Investigation Team)*

17

*Fig. 1.21. Although no damage is apparent to the main building, the amplified response at roof level has damaged the cross, in the Mexican earthquake, 1985 (photograph by courtesy of C. Taylor, Earthquake Engineering Field Investigation Team)*

*Fig. 1.22. Loose items are easily shaken off their supports, and this would easily be avoided by positive fixings: where they form part of an emergency power supply to a hospital lives may depend on a little engineering common sense (photograph by courtesy of J. Meehan)*

18

*Fig. 1.23. In the Mexico earthquake, 1985, one apartment block out of three identical structures failed completely: minor variations in ground conditions and construction quality may make the crucial difference (photograph by courtesy of E. Booth, Earthquake Engineering Field Investigation Team)*

*Fig. 1.24. Although there has been a catastrophic soft ground failure at one end of this building in Mexico, 1985, the framing system was sufficiently tough to survive without collapse: the structure would undoubtedly have to be demolished but the occupants would have been able to survive (photograph by courtesy of E. Booth, Earthquake Engineering Field Investigation Team)*

*Fig. 1.25. This Mexican reinforced concrete frame structure collapsed in the 1985 earthquake: the stair tower with a central reinforced concrete shear wall survived despite being monolithically connected to the rest of the building (photograph by courtesy of M. Winney)*

the capacity of the mounting so that it fails. Stand-by generators, air conditioning compressors and air handling fans are particularly vulnerable. Elevators are frequently put out of action. This can be due to damage to the headgear machinery or to displacement of the counter-weights from their tracks. Brittle types of pipework such as cast iron are liable to damage.

Emergency equipment is often mounted in racks and is unavailable because the racks have toppled or the equipment has fallen out. Stand-by batteries are often rack mounted and there are numerous instances where they have fallen out just when they were needed.

### 1.14. Conclusions

Lessons can and must be learned from earthquake damage. No proper feel or understanding of a building in earthquake motion can be acquired without looking at damage studies.

In the case of machinery a data base is being built up of experience with different categories (Anand, 1985). For buildings this is not possible because each structure is unique and has to be considered as a problem on its own (Figs 1.23–1.25).

### 1.15. Bibliography

Earthquake reconnaissance reports are prepared and published by the Earthquake Engineering Research Institute, 366–40th Street, Oakland, California 94609, USA.

Numerous damage studies appear in the proceedings of international conferences, in particular

(a) the four-yearly World Conference on Earthquake Engineering, held by the International Association for Earthquake Engineering, Kenchiku Kaikan 3rd Floor, 5-26-20, Shiba, Minato-ku, Tokyo 108, Japan

(b) the four-yearly European Conference on Earthquake Engineering held by the European Association for Earthquake Engineering, Central Laboratory for Earthquake Engineering, Bulgarian National Academy for Earthquake Engineering, Acad. G. Bonchev Str., Bl. 3, Sofia 1113, Bulgaria.

# Chapter 2

# Ground motion

'What to do in an earthquake? Stand still
and count to one hundred. By then it
will be over.' Professor G. W. Housner,
speaking on the BBC Overseas
Service, 1972

The scope of this chapter covers

  (a) the nature of ground motion
  (b) the principal factors in assessing ground motion
  (c) influences on ground motion
  (d) means of describing ground motion
  (e) the design earthquake.

## 2.1. Design objectives

The first objective is to consider the risk to the site itself from large
soil movements. These may be due to consolidation, liquefaction (the
temporary loss of shear strength), landslides or avalanches. Where
the earthquake fault intersects the ground surface large shear displacements may occur. Coastal sites would also need to consider tsu-
namis (commonly referred to as tidal waves).

The second objective is to define the nature of ground motion to be
expected on the site. Here problems in defining the earthquake
parameters and in relating them to more than one level of building
performance are encountered.

There is no single measure of an earthquake's capacity to cause
damage. Peak acceleration is the most commonly used parameter, but
peak velocity, frequency distribution and duration are also important. If one or more parameters are selected in order to characterise
an earthquake, a frequency of occurrence can be assigned to it in two
ways. If an average return period is calculated, say 50 years, then an
earthquake of that strength or greater can be expected 20 times in a
1000 year period. Alternatively a probability of occurrence can also
be calculated. If the average return period is $R$, then the probability
of occurrence in $y$ years is

$$P_y = 1 - \left(1 - \frac{1}{R}\right)^y \tag{2.1}$$

For example, for a 50 year return period, the probability of
occurrence in a 50 year period is $1 - 0.98^{50} = 0.64$.

More than one level of design may need to be considered. Typical levels are

(a) minor earthquake: elastic response—no damage
(b) moderate earthquake: inelastic response—minor structural damage and some non-structural damage
(c) major earthquake: substantial inelastic response without collapse—major structural and non-structural damage.

For major industrial and nuclear installations more specialised criteria may be used. The operating basis earthquake is one for which the installation should continue in operation, and the safe shut-down earthquake is one for which shut-down is acceptable but damage to critical facilities is not. In this context critical facilities are those where damage could lead to the release of dangerous chemicals or radioactivity.

## 2.2. Earthquake mechanics

### 2.2.1. Terminology

Figure 2.1 shows the principal terms used in describing an earthquake.

The source mechanism consists of the slipping of large masses of rock along a fault plane. Although the slip may occur over a large area it is commonly regarded as emanating from a single source called the focus.

### 2.2.2. Earthquake scales

There are two principal measures of an earthquake. The disturbance at source is measured by *magnitude* on a scale originally devised by C. F. Richter. Commonly used magnitudes, which are calculated from seismograph records, are $M_L$, the local magnitude, $M_s$, the surface wave magnitude, $m_b$, the body wave magnitude, and $M_w$, the moment magnitude. The existence of these different measures is

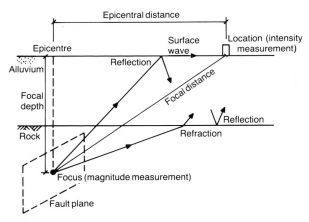

*Fig. 2.1. Earthquake transmission*

confusing and for a discussion of their uses Housner & Jennings (1982) is a good guide. Magnitudes can go up to 8·9 which is the highest value which has so far been recorded. The scales are not linear—one unit on the scale represents approximately a 32 fold increase in energy release at source.

The second measure is that of *intensity* which defines the effect of the earthquake at a point on the earth's surface. Intensity values are derived from subjective factors such as types of building damage, human perception, whether or not church bells ring etc. Several scales exist, and one of these, the modified Mercalli scale, is shown in Table 2.1. The need for intensity scales exists to assign values to historical events and to provide a simple method of describing the level of damage from a particular event.

### 2.2.3. Earthquake transmission

Two types of wave, P (compression) and S (shear), are generated by the source fault and radiate from it, each at its own characteristic speed. At the surface, secondary Love and Rayleigh waves are also generated. The waves are refracted and reflected at discontinuities and reflected at the surface, so that at any point a complex combination of wave-forms arrives over a period greater than that taken for the original disturbance. Newmark & Rosenblueth (1971) provide a guide to the generation and characteristics of strong ground motion.

### 2.2.4. Influences on ground motion

In spite of the random nature of earthquake motion on the surface, certain general trends relating distance, magnitude and period emerge.

Figure 2.2 shows the relationships between predominant period, distance and magnitude. There is a widespread belief that the pre-

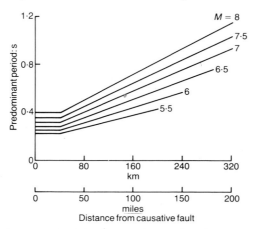

*Fig. 2.2. Predominant period–distance relationship for the maximum acceleration in rock (after Seed (1968))*

*Table 2.1.   Modified Mercalli scale of earthquake intensity*

I     Not felt

II     Felt by persons at rest, on upper floors, or favourably placed

III     Felt indoors; hanging objects swing; vibration like passing of light trucks; duration estimated; may not be recognised as an earthquake

IV     Hanging objects swing; vibration like passing of heavy trucks, or sensation of a jolt like a heavy ball striking the walls; standing motor cars rock; windows, dishes, doors rattle; glasses clink; crockery clashes; in the upper range wooden walls and frames creak

V     Felt outdoors, direction estimated; sleepers wakened; liquids disturbed, some spilled; small unstable objects displaced or upset; doors swing; shutters, pictures move; pendulum clocks stop, start, change rate

VI     Felt by all; many frightened and run outdoors; persons walk unsteadily; windows, dishes, glassware broken; small objects, books fall off shelves and pictures fall off walls; furniture moved or overturned; weak plaster or masonry D cracked; small bells ring (church, school); trees, bushes shaken

VII     Difficult to stand; noticed by drivers of motor cars; hanging objects quiver; furniture broken; damage to masonry D including cracks; weak chimneys broken at roof line; fall of plaster, loose bricks, stones, tiles, cornices; some cracks in masonry C; waves on ponds, water turbid with mud; small slides and caving in along sand or gravel banks; large bells ring; concrete irrigation ditches damaged

VIII     Steering of motor cars affected; damage to masonry C, partial collapse; some damage to masonry B, no damage to masonry A; fall of stucco and some masonry walls; twisting, fall of chimneys, factory stacks, monuments, towers, elevated tanks; frame houses moved on foundations if not bolted down, loose panel walls thrown out; decayed piling broken off; branches broken from trees; changes in flow or temperature of springs and wells; cracks in wet ground and on steep slopes

IX     General panic; masonry D destroyed, masonry C heavily damaged, sometimes with complete collapse, masonry B seriously damaged; frame structures, if not bolted, shift off foundations; frames racked; serious damage to reservoirs; underground pipes broken; conspicuous cracks in ground; in alluvial areas sand and mud ejected, earthquake fountains, sand craters

X     Most masonry and frame structures destroyed with their foundations; some well-built wooden structures and bridges destroyed; serious damage to dams, dikes and embankments; large landslides; water thrown on banks of canals, rivers, lakes etc; sand and mud shifted horizontally on beaches and flat land; rails bent slightly

XI     Rails bent greatly; underground pipelines completely out of service

XII     Damage nearly total; large rock masses displaced; lines of sight and level distorted; objects thrown into the air

*Table 2.1—continued*

| | |
|---|---|
| **Masonry categories** | |
| Masonry A | Good workmanship, mortar and design; reinforced, especially laterally, and bound together using steel, concrete etc.; designed to resist lateral forces |
| Masonry B | Good workmanship and mortar; reinforced but not designed in detail to resist lateral forces |
| Masonry C | Ordinary workmanship and mortar; no extreme weaknesses like failing to tie in at corners, but neither reinforced nor designed against horizontal forces |
| Masonry D | Weak materials such as adobe; poor mortar; low standards of workmanship; weak horizontally |

dominant period for any ground motion is in the range 0·25–0·3 s. This is due to the fact that a great deal of study has been done on Californian earthquakes which are typically shallow—25 km or so deep—and magnitude 5·0–6·5. In the Caribbean for example, where focal depths are typically near to 100 km, periods are substantially longer, at 0·35–0·4 s. Fig. 2.3 shows another relationship, this time between peak acceleration, magnitude and distance.

Figures 2.2 and 2.3 should only be taken as indicating trends, not exact relationships. The smooth curves show median values derived

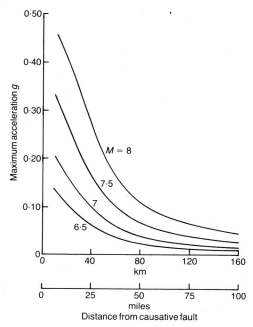

*Fig. 2.3. Acceleration–magnitude–distance relationship (after Seed et al. (1976))*

from well-scattered points on a graph. Subsoil properties also influence ground motion and this is dealt with in Section 2.5.

## 2.3. World seismicity

The outer skin of the earth, the lithosphere, which is strong and brittle, overlays a soft viscous layer, the athenosphere. The distribution of earthquakes, illustrated in Fig. 2.4, follows clearly defined zones which are characteristically associated with major geographical features such as the edges of continents and island arcs.

The lithosphere comprises a number of separate rigid plates, moving slowly but steadily relative to each other, and earthquakes tend to occur at the highly stressed plate boundaries. Various types of boundary movement occur, each with its own characteristic pattern of seismic and volcanic activity. Although earthquake occurrence is strongly concentrated at the plate boundaries there are many areas of seismic activity within the plates and remote from the boundaries.

## 2.4. Regional seismicity

### 2.4.1. Historical records

Catalogues of seismograph records of earthquakes world-wide have been published since 1918, whereas historical records go back to as early as 1200 BC in some parts of the world. For this reason the collection and analysis of ancient records forms an important part of our knowledge of seismicity. Geological studies also provide important clues to past activity, especially where earthquake sources are shallow and faults can be identified on the earth's surface.

Sufficient historical records can be interpreted to give frequencies of earthquakes of a given intensity at any point. Seismograph records give magnitudes and depths and can be used to give frequencies of events of a specified magnitude. Geological studies will locate active faults near which earthquake foci are likely to lie.

In assessing the seismicity of a site all the available information needs to be considered. Once this has been done, it should be remembered that any seismicity prediction remains an estimate with a substantial degree of uncertainty. Detailed assessments of the seismic hazard or of the properties of the design earthquake are regarded as being outside the scope of this work and are the province of geologists and engineering seismologists. Suitable references relating to this subject are given at the end of the chapter.

### 2.4.2. Faults

Regional studies of earthquakes show that the foci tend to lie on planes of weakness, or active faults. The extent of these faults can be identified both from seismograph records of earthquakes and from geological studies, especially where the faults extend to the earth's surface. Faults are not necessarily active. For many known geological faults the disruptive process that caused them has died down.

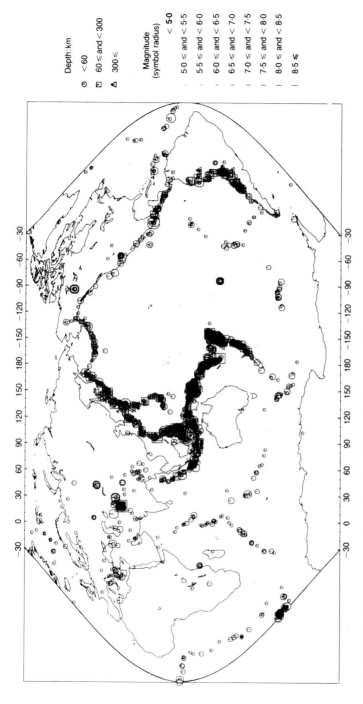

Depth: km

⊖ < 60

⊞ 60 ≤ and < 300

◁ 300 ≤

Magnitude
(symbol radius)

· < 5·0

· 5·0 ≤ and < 5·5

· 5·5 ≤ and < 6·0

· 6·0 ≤ and < 6·5

· 6·5 ≤ and < 7·0

| 7·0 ≤ and < 7·5

| 7·5 ≤ and < 8·0

| 8·0 ≤ and < 8·5

| 8·5 ≤

*Fig. 2.4. World occurrence of earthquakes of magnitude greater than 5.0; 1984–85 (prepared by the British Geological Survey, Edinburgh)*

29

### 2.4.3. Ground motion records

There are two principal types of ground motion record. Seismograph records are made continuously on a world-wide network and are used primarily for studies of earthquake magnitudes and location. Seismographs are very sensitive instruments for measuring displacement so that strong ground motions take the record off the scale and peak values are often not recorded.

Strong motion accelerographs are set into action by the earthquake itself and record the acceleration–time plot for the whole event, excluding the initial build-up to the level needed to trigger the instrument. Displacement and velocity records can be derived from the accelerograph with reasonable accuracy. Digitised records of earthquakes are available from a number of sources.

Because the duration of an earthquake is short—60 s is a long record—strong motion records contain little information about the very low frequency components of ground motion. It should also be borne in mind that strong motion records are commonly taken in locations in or near heavy buildings or engineering structures which have some filtering effect, biasing the frequency content.

### 2.4.4. Risk and uncertainty

As long as it is either not possible to estimate the most damaging earthquake which can occur or economically impracticable to design for the extreme event a small but finite risk of failure will exist. An idea of the position of earthquakes in the scheme of natural events is shown in Fig. 2.5. For the relatively frequent event, death from earthquakes is less than for other natural disasters, but for the rare event earthquakes take over as the chief destroyer.

Although analytical methods are well established by which the risk of occurrence can be calculated, such estimates are only as good as the available data.

### 2.4.5. Areas of low seismicity

The public is, naturally enough, greatly concerned with areas of high seismicity. In many areas of low seismicity the problem can be dismissed too readily. Few areas of the world are free from earthquakes altogether.

A normal well-constructed building, designed for moderate wind forces, should be well able to resist minor ground shaking. However, unusual structures which are flexible, or possibly brittle, may be sensitive to small nearby earthquakes or more distant larger earthquakes. The importance of earthquake resistance is emphasised for structures of great importance or where the consequences of failure are especially serious—such as nuclear reactors or stores for dangerous chemicals. The design earthquake for a sensitive structure in an area of low seismicity is most likely to be a comparatively small magnitude event occurring close to the site. The ground motion resulting from this characteristically has a high peak acceleration but a short duration.

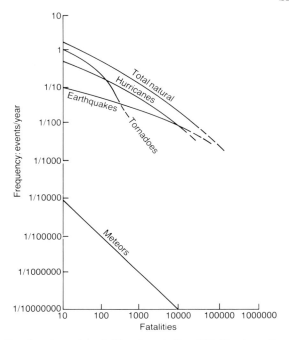

*Fig. 2.5. Deaths due to natural disasters (after US Nuclear Regulatory Commission (1976))*

## 2.5. Site effects

### 2.5.1. Caracas earthquake, 1967

Caracas, the capital city of Venezuela, suffered a severe earthquake on 29 July 1967 which totally destroyed a number of high-rise buildings. All the buildings of over 14 storeys which collapsed in the earthquake were in a single suburban area, Los Palos Grandes, which lies close to the deepest layer of alluvium underlying the city. It was well known before this that the dynamic properties of the soil underlying the site affected the nature of ground motion at the surface. However, the Caracas earthquake provided a powerful illustration of the critical importance of local geology.

Subsequent studies by Seed, Idriss & Dezfulian (1970) and others showed very clear relationships between the depth of alluvium, to which was attributed a fundamental natural period, and the type of structure that was most affected. Damage levels are high, as would be expected when the natural period of vibration for the soil and building are close, and this is illustrated in Fig. 2.6.

Figure 2.7 shows one dynamic model used to analyse the response of a number of layers of subsoil. Assuming accurate estimates can be made of the soil mass, damping and stiffness, established computer programs can relate surface motion to bedrock motion with reasonable accuracy.

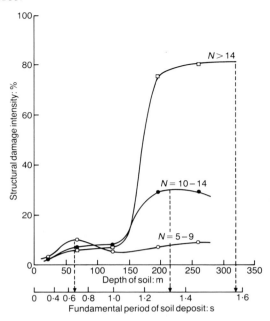

Fig. 2.6. *Relationship between damage and the natural period of the soil in the 1967 Caracas earthquake (after Seed (1970)).* $N$ = *no. of storeys;* $N/10 \simeq$ *natural period of building.*

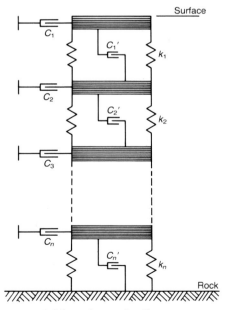

Fig. 2.7. *Lumped mass model for a layered soil*

General indications of the effect of soil on surface motion can be seen from Figs 2.8 and 2.9. Fig. 2.8 shows the relationship between alluvium depth and soil period, and Fig. 2.9 shows a relationship between peak acceleration and soil type. Both these relationships are general only and the scatter on each side of the best fit line is considerable.

The relationships in Fig. 2.9 are important. For low amplitude shaking, quite large amplifications are possible in very soft soil—for instance amplifications of over 20 were recorded for San Francisco Bay mud. However, this effect is swiftly overcome by yielding of soft soils as amplitudes increase, so that, for strong shaking, peak accelerations are normally reduced in transmission through the upper soils.

### 2.5.2. *Soil–structure interaction*

Considering the dynamic system shown in Fig. 2.7 it is clear that surface responses will be modified if another structure is added at the upper level. This appears to add yet another complication to the task of assessing a design input, especially if the large literature on soil–structure interaction is encountered. However, typical buildings have insufficient mass or stiffness to make a significant change in surface motion. The areas of concern (which account for much of the extensive literature) are those involving very heavy, rigid structures such as nuclear containment buildings.

An aspect of this problem which may be of more concern is that the connection of the building to the subsoil is not rigid, and local elastic deformation around the foundation can affect the dynamic modelling of the structure. This is discussed further in Chapter 3.

## 2.6.  Representation of ground motion

Figure 2.10 shows plots of acceleration, velocity and displacement against time. The acceleration plot is a record from a strong motion

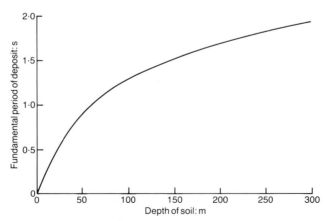

*Fig. 2.8. Relationship between the natural period of the soil and alluvium depth (after Seed (1970))*

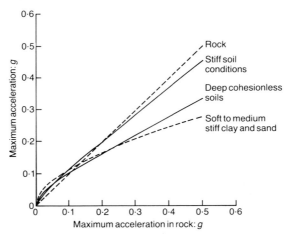

Fig. 2.9. *Effect of local soil conditions on peak acceleration (after Seed et al. (1976)): the relationships shown above 0·3g are based on an extrapolation of the data base*

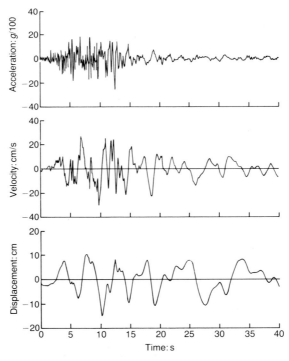

Fig. 2.10. *Time-based earthquake strong motion records (from Earthquake Engineering Research Laboratory (1980))*

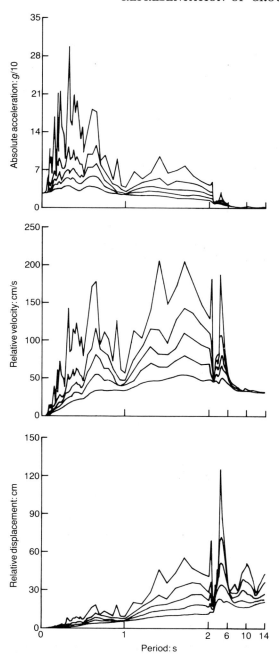

*Fig. 2.11. Response spectra from recorded strong motion records (from Earth-quake Engineering Research Laboratory (1980)) (San Fernando earthquake, 9 February 1971; damping values are 0%, 2%, 5%, 10% and 20% of critical)*

accelerograph and the other two plots have been obtained from it by integration. This is the simplest type of record and provides precise information about one specific earthquake. Important parameters associated with time-based records are the peak values and the duration of strong motion. For computation, earthquake records with digitised values at intervals of 0·02 s or 0·01 s are commonly used.

As an alternative to time-based records, frequency-based plots, or spectra, are used. Each spectrum represent a family of time-based records. There are a large number of types of spectrum, but those in common use are Fourier, response and power spectra.

The *response spectrum* represents the peak displacement, velocity or acceleration of a single degree of freedom spring–mass–damper system to an earthquake, plotted against its natural frequency. Continuous plots are drawn for each value of damping selected so that a response spectrum is represented by a family of curves, as in Fig. 2.11. The use of the response spectrum is described in Chapter 3.

Although there is a unique response spectrum for each time history, the reverse is not true. There is an infinite number of time histories that are compatible with each response spectrum.

The *Fourier spectrum* gives the frequency content of a record obtained by Fourier analysis, the values plotted being the coefficients $A_p$ in equation (2.2).

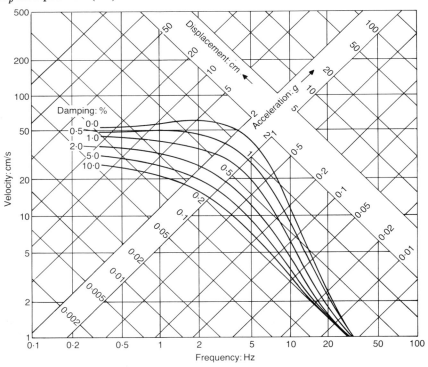

Fig. 2.12. Design response spectrum (after Housner (1970))

$$x(t) = \text{Re}\left[\sum_{p=1}^{\infty} A_p \exp(\mathrm{i}p\omega_0 t)\right] \qquad (2.2)$$

where $x(t)$ is a time varying function, Re is the real part of the expression, $\omega_0$ is the angular frequency and $t$ is the time.

For a full discussion of Fourier spectra and power spectra the reader is referred to Newland (1975).

The *power spectrum*, also referred to as the power spectral density, may be derived directly from the Fourier spectrum, being the values of $A_p/\omega_0^2$.

Spectra derived from a specific time-based record are, like the response spectrum in Fig. 2.11, rather spiky, or raw. For design purposes spectra are commonly averaged over a number of records so that they can be presented as smooth curves such as the design response spectrum in Fig. 2.12. In using smooth spectra it should be remembered that values may be distributed both above and below the curve. Design spectra are often plotted as plus one or two standard deviations, representing lines containing 84% and 98% of values respectively.

The smooth earthquake power spectrum can be represented by the Kanai–Taijimi expression (Kanai, 1967)

$$G(\omega) = \frac{G_0[1 + 4\xi_g^2(\omega/\omega_g)^2]}{[1 - (\omega/\omega_g)^2]^2 + 4\xi_g^2(\omega/\omega_g)^2} \qquad (2.3)$$

where $G(\omega)$ is the power spectrum, $\omega_g$ is the natural ground frequency, $\xi_g$ is the viscous damping for the ground and $G_0$ is the intensity parameter.

Values for $G_0$ may be derived from equation (2.4) (Section 2.7.4).

### 2.6.1. Damaging capacity
There is no generally accepted measure of the capacity of a specific earthquake to cause damage. Peak acceleration is unreliable because it may occur as only the briefest of transient values. The design earthquake is commonly characterised by the 'effective peak acceleration' and the 'effective peak velocity'. These represent bounding values for typical ground motion response spectra over the frequency ranges of interest in building design. Their use and derivation is described by the Building Seismic Safety Council (1985).

## 2.7. Design earthquake
### 2.7.1. Codes and equivalent static forces
Most building structures are designed on the basis of an earthquake design code and make no direct use of ground motion either as a spectrum or a time-based record. Instead, a set of static forces is applied which leads to a design providing adequate earthquake resistance (this subject is dealt with in Chapter 6). The resulting method produces a design approach that is suitable for the non-specialist but is not applicable to buildings of irregular form, irregular mass distribution or of unusual concept. Furthermore the static force

approach provides no information on behaviour during an earth-quake—and this may be needed where the contents are valuable or sensitive to vibration.

Thus there is a significant proportion of all buildings for which a design ground motion needs to be defined. As stated earlier the definition of such a motion is outside the scope of this work; a discussion of the factors involved is given by Seed & Idriss (1982). Analysis is discussed in Chapter 3 and the approach selected will principally determine which definition of ground motion is adopted.

### 2.7.2. Time-based records

Two earthquakes with identical peak acceleration will produce different displacements and forces in a building structure. Hence several time-based records will be required to give an estimate of the mean response and the distribution of values above and below the mean. Recorded earthquakes can be scaled to identical peak acceleration in order to give some measure of consistency, but it is difficult to select records with similar spectral properties and durations.

An alternative approach, where an estimate of the spectral properties can be made, is to use computer-generated time-based records. These can be produced to fit a defined spectrum, time envelope and peak acceleration. Programs are available to do this, some of which are listed in Appendix 1.

### 2.7.3. Response spectrum

Elastic response design spectra can be produced for differing site conditions and scaled for an appropriate peak acceleration. A typical example is shown in Fig. 2.13 which is for 5% structural damping.

Response spectra are influenced by factors other than soil type. Recommended spectra taking account of the distance of the source are available, but the present lack of adequate data does not permit specific recommendations to take account of magnitude or source mechanism.

### 2.7.4. Power spectrum

Power spectra can be derived from time-based records by Fourier analysis of the strong motion segment. More commonly a design power spectrum is derived from equation (2.3) with suitable values inserted for $G_0$, $\omega_g$ and $\xi_g$.

Suggested values (Kanai, 1967) for firm ground are $\omega_g = 12.7$ and $\xi_g = 0.6$. $G_0$ is a scaling factor and can be related approximately to the peak value by

$$G_0 = \frac{0.141 a_{\max}{}^2 \xi_g}{\omega_g (1 + 4\xi_g{}^2)^{1/2}} \tag{2.4}$$

where $a_{\max}$ is the peak value of the time-based record.

### 2.7.5. Vertical, torsional and rocking components

For most structures the horizontal component of earthquake

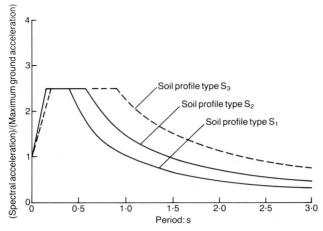

*Fig. 2.13. Recommended design spectrum for varying soil types (after ATC3-06): soil profile type $S_1$—rock of any characteristic, either shale like or crystalline in nature (such material may be characterised by a shear wave velocity greater than 2500 ft/s) or stiff soil conditions where the soil depth is less than 200 ft and the soil types overlying rock are stable deposits of sands, gravels or stiffer clays; soil profile type $S_2$—deep cohesionless or stiff clay soil conditions, including sites where the soil depth exceeds 200 ft and the soil types overlying rock are stable deposits of sands, gravels or stiff clays; soil profile type $S_3$—soft–medium stiff clays and sands, characterised by 30 ft or more of soft–medium stiff clay with or without intervening layers of sand or other cohesionless soils*

ground motion has by far the greatest effect. However, the other components should not be forgotten and may need to be explicitly taken into account in some cases.

Vertical motions affect long span structures and must be allowed for in stability calculations. The customary assumption is that the peak vertical acceleration is two-thirds of the peak horizontal acceleration and has a similar spectral distribution. However, this is a crude approximation and a more accurate assessment needs to be made for structures that are sensitive to vertical ground shaking.

Torsional motion is usually allowed for by assuming that the horizontal motion is applied eccentrically at the base of the building. A typical code requirement is to apply an eccentricity of 5% of the base dimension of the building to the base motion to account for torsional motion, unforeseen eccentricity in the building and the possible coupling of torsional and lateral response. Rocking components are seldom used as design criteria, the effects being similar to those from lateral motion in most respects.

Rutenberg & Heidebrecht (1985) give simple approaches to the construction of response spectra for torsion and rocking. Hutchinson & Chandler (1986) provide a method for dealing with the coupling of torsional and translational modes.

## 2.8. Bibliography

*2.8.1. General*

Bolt, B. A. (1978). *Earthquakes.* New York: Freeman.

Housner, G. W. & Jennings, P. C. (1982). *Earthquake design criteria.* Berkeley: Earthquake Engineering Research Center.

Lomnitz C. & Rosenblueth, E. (1976). *Earthquake risk and engineering decisions.* Amsterdam: Elsevier.

Seed, H. B. & Idriss, I. M. (1982). *Ground motion and soil liquefaction during earthquakes.* Berkeley: Earthquake Engineering Research Institute.

Wiegel, R. L. (1970). *Earthquake engineering.* Englewood Cliffs: Prentice Hall.

*2.8.2. Information on earthquake occurrence and ground motion records*

British Geological Survey, Murchison House, West Mains Road, Edinburgh EH9 3LA, Scotland.

Department of Civil Engineering, Imperial College of Science and Technology, London, England.

International Seismology Centre, Newbury, Berkshire RG13 1LX, England.

National Geophysical Data Centre, NOAA, 325 Broadway, Boulder, Colorado 80303, USA.

National Information Service for Earthquake Engineering, Earthquake Engineering Research Center, University of California, 451RFS, 47th Street and Hoffman Boulevard, Richmond, California 94804, USA.

United States Geological Survey, Golden, Colorado, USA.

*Chapter 3*

# The calculation of structural response

CALCULATE v.t., to ascertain
beforehand by mathematical process.
*Cassells English dictionary*

The scope of this chapter covers

(*a*) single degree of freedom response
(*b*) multiple degree of freedom response
(*c*) deterministic linear response
(*d*) probabilistic linear response
(*e*) deterministic non-linear response
(*f*) probabilistic non-linear response
(*g*) ductility requirement
(*h*) soil–structure interaction
(*i*) secondary structure response
(*j*) capacity design
(*k*) practical analysis procedures
(*l*) notes on structural modelling
(*m*) examples.

## 3.1. Design objectives

Information on the forces in structural elements is required, corresponding to a particular level of earthquake, and acceptable levels of structural and non-structural damage. Normally accepted levels are given in Table 3.1.

Although structural response to a major earthquake is inelastic, and many members will have yielded and developed plastic hinges, non-linear analytical procedures are not used as commonly as might be expected. In practice a great deal of dynamic analysis is carried out on the basis of elastic response on the assumption that sufficient ductility and redundancy will be provided in the structure to accommodate the non-linear response to a major earthquake. The design philosophy that accompanies this approach is described in detail in Chapter 6.

## 3.2. Response

The response of a structure to an earthquake may refer to stress, displacement, acceleration, velocity, shear or any other parameter

Table 3.1. *Acceptable levels of damage*

| Earthquake | Structural damage | Non-structural damage |
|---|---|---|
| Minor | None | None |
| Moderate | None | Some |
| Major | Some but not collapse | Substantial |

affected by ground motion. Response may be defined in time, but it is customary to refer to the response as the peak value of the particular parameter caused by the earthquake.

The objectives of the dynamic analysis of a structure responding to dynamic forces can be

(a) to establish strength and ductility requirements
(b) to calculate equivalent static forces for design
(c) to calculate displacements
(d) to establish the nature of dynamic design input to equipment mounted on the structure—machinery, pipework, storage tanks etc.

In general, the effect of vibration on a structure can be calculated if a numerical model of both the vibration, or 'forcing function', and the structure can be arrived at. Models for ground motion are discussed in Chapter 2, but the actual values may be modified by soil–structure interaction which is discussed further in Section 3.10. Ground motion is commonly defined in terms of two horizontal components, although it consists in reality of these plus one vertical, two rocking and one torsional component. The reasoning behind this is that by far the greatest effect on normal buildings is caused by the horizontal components.

Where the building is eccentric on plan, i.e. the centres of mass and stiffness are substantially separated, the effect of torsion may be sufficiently large to cause significant increases in lateral displacements. This effect is greatest on the outer elements of the structure. The coupling of response to torsional and lateral components may be high when the natural frequencies of the two modes are close.

The linear structural model requires information on structural stiffness, derived from the geometric and material properties in the same manner as for static analysis. In addition to this, because dynamic forces derive from the inertia of the structural masses, information on the mass distribution is required, and this can be either in the form of lumped masses at the structural nodes or as distributed mass. Fig. 3.1 shows some structural models using both finite elements and spring–mass systems. Buildings are customarily modelled as lumped mass models, distributed mass models normally being applied only to the analysis of structural elements.

## 3.3.  Single degree of freedom response

Figure 3.2 shows a single degree of freedom model, which could be representative of a single-storey structure where the columns are

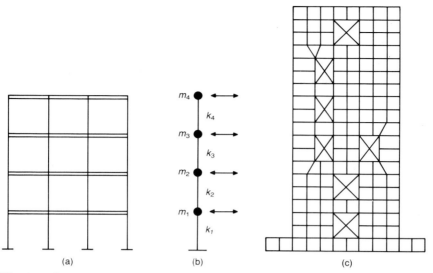

Fig. 3.1. *Structural models: (a) structural frame; (b) simplified stick model (translational degree of freedom for each mass; $m_i$ is the mass of the floor at level $i$; $k_i$ is the total column stiffness at level $i$): (c) finite element model of a shear wall*

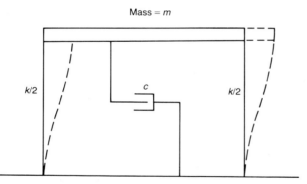

Fig. 3.2. *Single degree of freedom structure*

light in relation to the roof mass. The equation of motion is

$$\ddot{x} + 2\xi\omega\dot{x} + \omega^2 x = p(t)/m \qquad (3.1)$$

where $\omega = (k/m)^{1/2}$ is the natural circular frequency of vibration. The corresponding period of vibration $T = 2\pi/\omega$ and frequency $f = \omega/2\pi$.

For a harmonic force $p(t) = p_0 \sin(\bar{\omega}t)$ applied to the mass the steady state solution for the displacement $x$ is

$$x = \frac{x_0}{[(1 - \beta^2)^2 + (2\xi\beta)^2]^{1/2}} \qquad (3.2)$$

43

where $x_0$ is the displacement due to a static force $p_0$, $\xi$ is the damping ratio and $\beta$ is the ratio of the forcing frequency to the natural frequency, $\bar{\omega}/\omega$.

Figure 3.3 shows the effect of varying the frequency of excitation on the response. The maximum value of the response is

$$x \approx x_0/2\xi \tag{3.3}$$

When the frequency ratio is zero, this is equivalent to applying a static force so that the response ratio equals unity. For high frequencies the response ratio is approximately given by

$$x \approx x_0/\beta^2 \tag{3.4}$$

Damping values $\xi$ are given as a fraction of the critical damping, defined as the lowest value at which the structure does not vibrate. The damped natural frequency is given by

$$\omega_d = \omega(1 - \xi^2)^{1/2} \tag{3.5}$$

which for normal structural damping values (less than 0·1) means that the damped and undamped natural frequencies are approximately equal.

For earthquake motion, if displacement is measured relative to the ground, the equation of motion becomes

$$\ddot{x} + 2\xi\omega\dot{x} + \omega^2 x = -\ddot{x}_g(t) \tag{3.6}$$

The solution of equation (3.6) is the Duhamel integral

$$x(t) = -\frac{1}{\omega_d} \int_0^t \ddot{x}_g(\tau) \exp[-\xi\omega(t - \tau)] \sin[\omega_d(t - \tau)] \, d\tau \tag{3.7}$$

which gives the displacement of the structure at any time $t$, and $\omega_d$ is defined by equation (3.5).

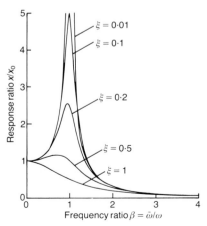

*Fig. 3.3. Single degree of freedom response*

Using equation (3.7) it is possible to obtain the response of a single degree of freedom system with natural frequency $\omega$ and damping $\xi$ at any time $t$, for a specific earthquake. For design purposes the peak absolute values during the earthquake are required, and these, when plotted, give the response spectrum. Fig. 3.4 shows a set of response spectra for the 1940 Imperial Valley earthquake. The quantities plotted are

(a) $S_d$, the spectral displacement
(b) $S_v$, the spectral pseudovelocity
(c) $S_a$, the spectral pseudoacceleration

The 'pseudo' prefix means that the quantity is derived from the energy content and is not precisely the same as the actual relative velocity or acceleration. Within the range of frequencies normally encountered in building design the distinction between pseudo and actual is not significant.

Response spectra may be given for a specific recorded earthquake motion or may be design spectra to be used for a specific site, soil type or area. Generally design spectra will be smoothed and will represent an estimate of response for a particular probability of occurrence.

Figure 3.5 shows a smooth design spectrum and Fig. 3.6 shows one of the type common in the nuclear industry, where the points A, B, C, and D represent control points. The spectra are all shown plotted on tripartite bases, i.e. where spectral displacement, velocity and acceleration are shown on the same plot, made possible by the relationships

$$S_a = \omega S_v = \omega^2 S_d \qquad (3.8)$$

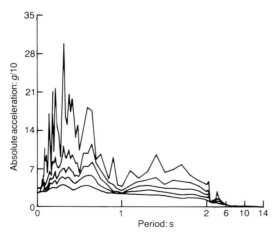

*Fig. 3.4. Acceleration strong motion spectrum (from Earthquake Engineering Research Laboratory (1980)) (San Fernando earthquake, 9 February 1971; damping values are 0%, 2%, 5%, 10% and 20% of critical)*

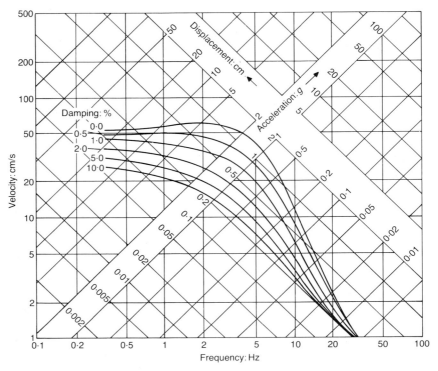

*Fig. 3.5. Design response spectrum (after Housner (1970))*

Figure 3.6 illustrates the typical structure of a response spectrum, consisting of distinct frequency zones

    (a) a low frequency zone where displacement is approximately constant (below D): at very low frequencies the displacement will eventually approach the peak ground displacement value, but this occurs at frequencies so low that they are out of the range encountered in normal buildings; the physical significance of this condition is that the building is effectively floating and unaffected by the ground motion

    (b) an intermediate zone (D–C), which embraces most of the typical first-mode frequencies of buildings, where the velocity is approximately constant

    (c) a third zone (C–B), where the acceleration is approximately constant

    (d) a high frequency zone (B–A), where the acceleration varies approximately linearly between the peak value and the peak ground acceleration

    (e) above A the peak building and ground accelerations are equal: the physical significance of this is that the high structure frequency implies that it is effectively rigid and simply reproduces

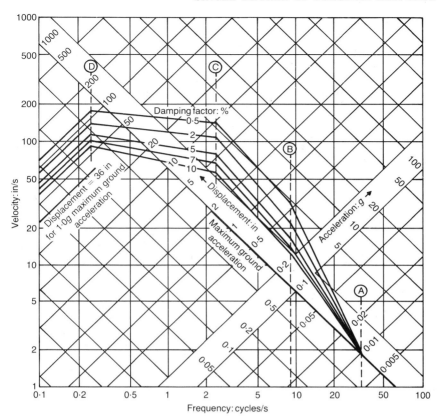

*Fig. 3.6. Design spectra for nuclear plant (after American Society of Civil Engineers (1980))*

the ground motion at all points; this value, sometimes referred to as the 'zero period acceleration' (ZPA), is often used as a scaling value for the whole spectrum, e.g. if a design spectrum has a ZPA of 1·0g, then values of displacement, velocity or acceleration taken from the spectrum for another ZPA, say 0·35g, are scaled by 0·35.

The response spectrum offers the simplest of tools to the designer of a single degree of freedom system. If the natural frequency and damping are known, the displacement, velocity and acceleration are read directly from the spectral plot. As the response of most buildings is mainly from their first mode, knowledge of the first-mode frequency will give a useful guide to the response.

For example for a 12 storey structure, with a damping of 5% responding to an earthquake with a ZPA of 0·35% the first-mode frequency can be roughly estimated from equation (3.1) as 0·87 Hz. From Fig. 3.6 this gives a peak displacement of approximately 18 in, corres-

ponding to the peak acceleration for which the spectra are plotted—
1·0$g$. Hence for a design earthquake of 0·35$g$ peak acceleration the
maximum displacement will be 0·35 × 18 in or 6·3 in.

### 3.4. Multiple degree of freedom response

Figure 3.7 shows the manner in which a set of mode shapes are
combined into the total geometric response of a multiple degree of
freedom system. The mode shapes are defined by the system eigenvectors, and their associated frequencies are derived from the eigenvalues. Both are derived from the eigensolution, which is the solution
of the equation

$$[K]\Phi = \{\omega_i^2\}[M]\Phi \qquad (3.9)$$

The matrix $\Phi$ represents the eigenvectors arranged columnwise and
the eigenvalues $a_1, a_2, \ldots a_n$ are equal to $\omega_1^2, \omega_2^2, \ldots \omega_n^2$. Solution of
the eigensystem is usually by computer but is also described by
Clough & Penzien (1975) and Bathe & Wilson (1976). For an $n$ degree
of freedom system there will be $n$ equations, $n$ mode shapes and $n$
modal frequencies.

The damped equations of motion for the multiple degree of freedom
system are

$$[M]\{\ddot{x}\} + [C]\{\dot{x}\} + [K]\{x\} = \{p(t)\} \qquad (3.10)$$

By the technique of normal coordinates $n$ separate equations can be
arrived at for each of the modes. For mode $j$

$$\ddot{Y}_j + 2\xi_j\omega_j\dot{Y}_j + \omega_j^2 Y_j = P_j(t)/M_j \qquad (3.11)$$

where

$$M_j = \{\phi_j\}^\mathrm{T}[M]\{\phi_j\} \qquad (3.12)$$

$$P_j(t) = \{\phi_j\}^\mathrm{T}\{p(t)\} \qquad (3.13)$$

and the total response (eqn. (3.14))

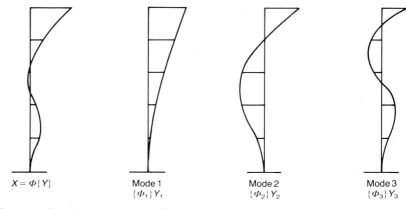

$$X = \Phi\{Y\} \qquad \qquad \text{Mode 1} \qquad \qquad \text{Mode 2} \qquad \qquad \text{Mode 3}$$
$$\{\Phi_1\}Y_1 \qquad \qquad \{\Phi_2\}Y_2 \qquad \qquad \{\Phi_3\}Y_3$$

*Fig. 3.7. Combination of modal responses*

$$\{x\} = \Phi\{Y\} \tag{3.14}$$

In general the first few modes will govern the response so that only these few will need to be summed in equation (3.14).

## 3.5. Deterministic linear analysis

Deterministic methods are those which result in specific answers, as opposed to the statistical answers produced by probabilistic methods. Perhaps the commonest and simplest approach under the deterministic heading is the *response spectrum* approach.

Equation (3.11) can be rewritten

$$\ddot{Y}_j + 2\xi_j\omega_j\dot{Y}_j + \omega_j^2 Y_j = \gamma_j\ddot{x}_{\mathrm{g}} \tag{3.15}$$

where the jth modal participation factor

$$\gamma_j = \frac{\{\phi_j\}^{\mathrm{T}}[M]}{\{\phi_j\}^{\mathrm{T}}[M]\{\phi_j\}} \tag{3.16}$$

Because the mode shapes are vectors of relative values they can be normalised. In order to simplify response calculations they are commonly normalised so that

$$\{\phi_j\}^{\mathrm{T}}[M]\{\phi_j\} = 1 \tag{3.17}$$

By referring to the appropriate displacement response spectrum, scaled to the peak value of $x_{\mathrm{g}}$, the modal amplitude for mode $j$ is

$$Y_j = \gamma_j S_{\mathrm{d}}(\omega_j) \tag{3.18}$$

and the vector of displacement response in mode $j$ is

$$\{x_j\} = \{\phi_j\}Y_j \tag{3.19}$$

Thus for each mode the peak displacements at each node are known. It is normally only necessary to sum a few modes as displacements in the higher modes tend to be very small. The summing of modal displacements is not arithmetical as the peak responses of the modes are unlikely to occur at the same time, although the sum will of course give an upper bound value. The common approach is to use the square root of the sum of the squares

$$x = \left(\sum_{j=1}^{n} x_j^2\right)^{1/2} \tag{3.20}$$

This method is reasonable where the modes are well separated in frequency. Where the separation is less than 10%, as defined by

$$\omega_j < 1{\cdot}1\omega_i \tag{3.21}$$

a simple approach is to use the square root of the sum of the squares method for the separated modes and the arithmetical sum for the closely spaced modes. A more comprehensive approach to closely spaced modes is given in Tsai (1984).

An example of using the response spectrum method is given in Section 3.15.

The second method of deterministic analysis is the *time history* method. This solves the equations of motion directly for small increments of time and traces the structure response through the whole period of the earthquake. The basic principles of the method are illustrated by the Newmark (1959) method, although alternatives exist such as the Wilson $\theta$ method (Clough & Penzien, 1975) and the Houbolt (1950) method.

The Newmark method uses the following relationships between acceleration, velocity and displacement at the beginning and end of a time interval $\Delta t$

$$\dot{x}_{t+\Delta t} = \dot{x}_t + \Delta t(1 - \delta)\ddot{x}_t + \Delta t\delta\ddot{x}_{t+\Delta t} \tag{3.22}$$

$$x_{t+\Delta t} = x_t + \Delta t\dot{x}_t + \Delta t^2(0\cdot5 - \alpha)\ddot{x}_t + \Delta t^2\alpha\ddot{x}_{t+\Delta t} \tag{3.23}$$

where $\delta$ and $\alpha$ are constants on which the accuracy and stability depend. For $\delta = 0\cdot5$ and $\alpha = 0\cdot1667$ the conditions of linear acceleration are produced and these are commonly used.

The equations of motion at time $t + \Delta t$ are

$$[M]\{\ddot{x}_{t+\Delta t}\} + [C]\{\dot{x}_{t+\Delta t}\} + [K]\{x_{t+\Delta t}\} = P_{t+\Delta t} \tag{3.24}$$

By combining equations (3.22)–(3.24), a step-by-step formulation is produced involving the solution at each time step of a set of equations of the form

$$[K^*]\{x_{t+\Delta t}\} = P_t^* \tag{3.25}$$

The matrix $[K^*]$ is not time dependent and may be triangularised at the beginning of the program. The vector $\{P^*\}$ is formulated at each time step and the equation solved for $\{x_{t+\Delta t}\}$.

The damping matrix $[C]$ in equation (3.24) may be obtained from (Clough & Penzien 1975)

$$[C] = [M]\Phi\zeta\Phi^T[M] \tag{3.26}$$

where

$$\zeta = \left[\frac{2\xi_n\omega_n}{m_n^*}\right]_{\text{diagonal}} \tag{3.27}$$

$$m_n^* = \{\phi_n\}^T\{m\}\{\phi_n\} \tag{3.28}$$

The time history method is only suitable for use on a computer and is relatively expensive to run, whereas the response spectrum method is cheap and the calculations for simple structures, once the eigensystem is known, may be done by hand. It is generally desirable to carry out at least three time history analyses for design purposes as considerable variations in response occur from one record to another, even when the records have the same peak acceleration; the example in Section 3.15 shows this.

Newmark (1959) provided maximum values for the time step based on the integration constants and the natural period of vibration of the structure, but generally smaller values are used in order to represent effectively the variations in the ground motion accelerogram. Customary values of time step are 0·02 s or 0·01 s for building structures.

### 3.6. Probabilistic linear response

Probabilistic response deals with 'stationary' vibrations. In this context stationary means that the statistical nature of the vibration—mean value, root-mean-square amplitude, frequency distribution—does not change with time. Examples of this are turbulent air flow around an aircraft flying at constant speed, noise in electrical circuits or ocean waves. Although earthquake motion is not stationary, during the time of strong motion a typical earthquake is approximately stationary, and good estimates of response can be made using random vibration theory. For lightly damped structures and those with a low natural frequency a correction factor can be applied to allow for the fact that the duration is limited.

The description of earthquake ground motion in the form of a power spectrum is given in Section 2.6 as

$$G(\omega) = \frac{G_0[1 + 4\xi_g^2(\omega/\omega_g)^2]}{[1 - (\omega/\omega_g)^2]^2 + 4\xi_g^2(\omega/\omega_g)^2} \qquad (3.29)$$

Power spectra and frequency response functions extend on both sides of the zero-frequency axis, about which they form mirror images, as shown in Fig. 3.8. The concept of negative frequencies is of no concern in structural response so that when integrating from $-\infty$ to $+\infty$ the integral is replaced with twice the integral from 0 to $+\infty$. Care should be taken in interpreting power spectra as they are sometimes given as one-sided spectra equal to the sum of the negative and positive sides.

The response of the structure can also be described in the frequency domain by the frequency response function $H_{x0}(\omega)$ equal to the response $x$ to a unit forcing function of frequency $\omega$. $H_{x0}(\omega)$ can be calculated from normal linear response theory. The output response function to a forcing function defined by $G(\omega)$ is then

$$H_x(\omega) = G(\omega)|H_{x0}(\omega)|^2 \qquad (3.30)$$

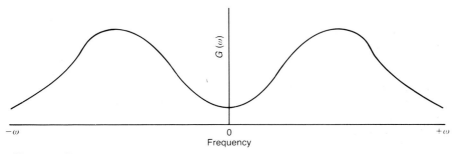

Fig. 3.8. Power spectrum

The root-mean-square response $\sigma_x$, given by

$$\sigma_x = \frac{1}{t} \sum_0^t x(t)^2 \, dt \tag{3.31}$$

is then defined by

$$\sigma_x^2 = \int_{-\infty}^{\infty} |H_x(\omega)| \, d\omega \tag{3.32}$$

Alternatively, ignoring negative frequencies

$$\sigma_x^2 = 2 \int_0^{\infty} |H_x(\omega)| \, d\omega \tag{3.33}$$

Allowance for the limited duration of strong motion may be made by substituting a modified damping value given by

$$\xi_t = \frac{\xi}{1 - \exp(-2\xi\omega_0 t)} \tag{3.34}$$

Vanmarcke (1976) gives more complex alternative approaches to the calculation of both peak response and the effect of limited duration.

Figure 3.9 illustrates the random vibration process. The root-mean-square response is derived from $H_x(\omega)$ by equations (3.33) and (3.34).

### 3.7. Deterministic non-linear response

Because earthquake design is concerned with structures that are strained well past the yield point, the analysis is non-linear. The first and simplest approach is to modify the linear method of response spectrum analysis, by constructing non-linear response spectra. A non-linear response spectrum differs from the linear version by using a yielding single degree of freedom oscillator in place of a linear oscillator. An additional parameter, ductility, is then introduced, representing the ratio of displacement to yield level displacement. An approach, described by Newmark & Hall (1982), enables non-linear response spectra to be plotted approximately from linear values for a given ductility factor. By using the response spectrum as the starting point the modified non-linear spectrum will represent the same damping and probability level. The Newmark method is applicable to earthquakes of normal duration.

Figure 3.10 shows a typical response spectrum consisting of lines of constant displacement (0–0·3 Hz), constant velocity (from 0·3 Hz to about 2 Hz) and constant acceleration (2–8 Hz). From 8 Hz to 33 Hz the line is transitional and above 33 Hz it follows the peak ground acceleration. To construct the displacement response spectrum line for a given ductility factor $\mu$ the procedure is as follows

(a) 0–0·3 Hz—inelastic response follows elastic values
(b) 0·3–2·0 Hz—inelastic velocity equals peak elastic velocity

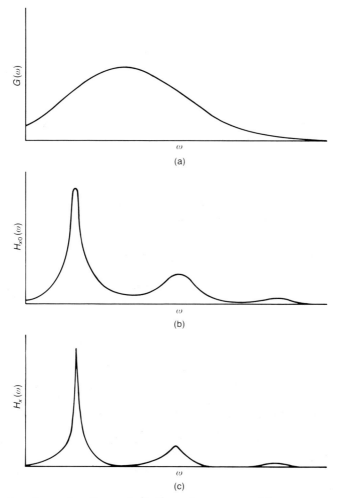

*Fig. 3.9. Random vibration calculation of response: (a) power spectrum of ground motion; (b) frequency response function; (c) output frequency response* $H_x(\omega) = G(\omega)\,|\,H_x(\omega)\,|^2$

(c) 2·0–8·0 Hz—inelastic acceleration equals elastic value times $k$ where

$$k = \frac{\mu}{(2\mu - 1)^{1/2}} \qquad (3.35)$$

(d) 8·0–33 Hz—transitional

(e) above 33 Hz—inelastic acceleration equals elastic acceleration times $\mu$.

Figure 3.11 shows elastic and inelastic spectra constructed by this method.

53

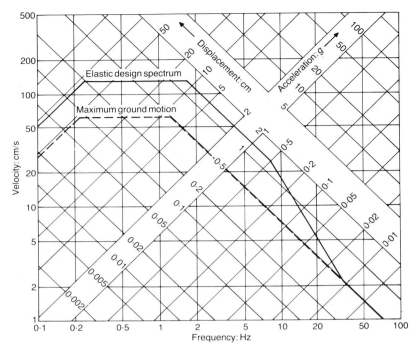

*Fig. 3.10. Typical design response spectrum*

In following the method it must be remembered that it is an approximation and should not be used for structures whose first-mode frequency is less than 1·0 Hz. The displacement response line plotted by this method only represents displacement correctly on the tripartite plot as the relationships given in equation (3.8) do not apply to non-linear response.

The alternative and much more accurate approach to non-linear structural analysis is the time history method. Because each time step is separately computed it is possible to review the structural parameters step by step and to modify them to take account of non-linear behaviour, either geometrical or material. Even such non-linear behaviour as impact or fracture can be modelled. However, the time interval required to ensure computational stability is much lower than that which can be used for linear time histories. Thus the expense of non-linear time domain analysis is much larger than the linear equivalent. A discussion of the various methods that are available is given by Adeli, Gere & Weaver (1978).

### 3.8. Probabilistic non-linear analysis

The effect of yielding in a single degree of freedom system is similar to that of increasing the damping and reducing the stiffness. A number of researchers have examined equivalent linear systems that reproduce the non-linear behaviour of yielding structural systems.

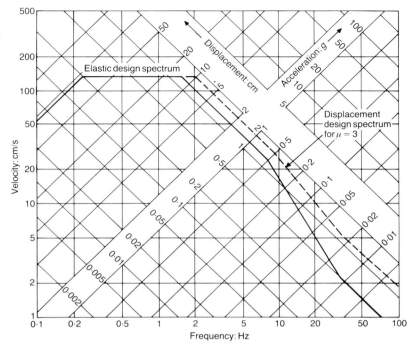

*Fig. 3.11. Elastic and inelastic design response spectra*

Several of these are discussed by Hadjian (1982) and he recommends that proposed by Iwan (1980), where the stiffness $k$ and damping $\xi$ are replaced by equivalent values $k_e$ and $\xi_e$ as follows

$$k_e = \frac{k}{[1 + 0\cdot121(\mu - 1)^{0.939}]^2} \qquad (3.36)$$

$$\xi_e = \xi + 0\cdot058(\mu - 1)^{0.371} \qquad (3.37)$$

An iterative process is required to match the ductility with that resulting from the particular excitation. Ideally it should be possible to replace the yielding section of a multiple degree of freedom system with the equivalent linear system, but the validity of this has not been established.

### 3.9. Ductility demand

Ductility of a member or structure is defined as the ratio of displacement at ultimate load to the displacement at yield. Structures have a limit to how much ductility they can provide, and some structural arrangements impose very high demands. A common example of this is where a small structure is attached to a large structure. In consequence it may sometimes be desirable to make an estimate of this requirement.

Non-linear time history analysis will provide this information, but, as has already been stated, this is an expensive process. An approximate method is described by Clough & Penzien (1975), based on the assumption that non-linear displacement is equal to linear displacement (a similar assumption was made in preparing non-linear response spectra up to 2 Hz). Fig. 3.12 illustrates the method, $f_{max}$ being arrived at by linear analysis.

Typical structure ductilities are of the order of 3–5. Member ductilities, based on curvature, that are compatible with these structural ductilities depend on the hinge mechanism which develops but are much higher. A discussion of the relationship between structure and member ductility is given in Park & Paulay (1975). Member ductility demand increases towards the top of a multi-storey building.

### 3.10. Soil–structure interaction

The interaction of the structure and its supporting soil falls into two categories. Buildings in general are light in relation to the mass of supporting soil and relatively flexible, so that the addition of the building does not affect the surface ground motion significantly. However, local flexibility of the soil where it is in contact with the foundation can modify the building's response.

The effects of this local flexibility are to modify vibrational modes, to lower natural frequencies and to generate additional damping through energy dissipation in the surrounding soil. Although an increase in response can occur, the general effect is to produce a reduction in base shear. Piled foundations, in comparison with bearing foundations, generally have a lesser effect on the mode shapes and frequencies but produce lower damping effects.

The modelling of local flexibility in the soil is dealt with in Chapter 9, together with the derivation of appropriate mechanical properties for the subsoil.

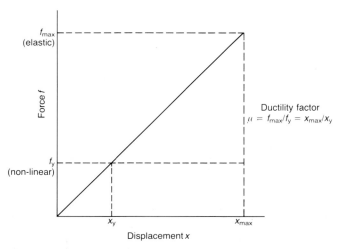

*Fig. 3.12. Ductility factor calculation*

The second type of soil–structure interaction to be considered is where a structure is massive and rigid. In this case it can affect the surface ground motion below and adjacent to its foundation. The structure modelling needs to incorporate the supporting soil layers down to a rock base. This type of analysis is a sophisticated and specialised matter and is only required for major, massive structures such as nuclear reactors. Fig. 3.13 shows models that are appropriate to the two types of soil–structure interaction. For a detailed work on the subject see Wolf (1985).

Although a full treatment of this second type of soil–structure interaction is outside the scope of this text, the complex frequency domain method of analysis which is commonly used in dealing with it is included here for completeness. Its particular advantage lies in dealing with soils where some of the properties are frequency dependent.

The method is deterministic, but instead of solving the equations of motion in the time domain the solution is carried out in the frequency domain. The Fourier transform pair are written thus

$$P(\omega) = \int_{-\infty}^{\infty} R(t) \exp(-i\omega t)\, dt \tag{3.38}$$

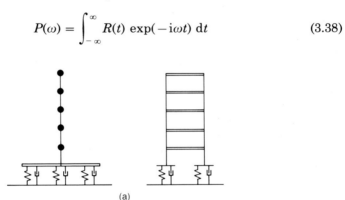

(a)

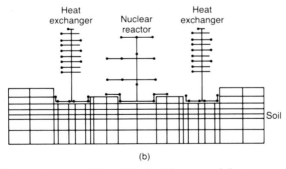

(b)

*Fig. 3.13. Soil–structure models: (a) building models to account for soil–foundation springing; (b) soil–structure model for a nuclear reactor (the nuclear reactor and heat exchanger models have been simplified; excitation is applied at the base of the model as a time history)*

$$R(t) = \frac{1}{2\pi} \int_{-\infty}^{\infty} P(\omega) \exp(i\omega t) \, d\omega \tag{3.39}$$

If the earthquake time history is assumed to be periodic, with a duration $T$, it can be represented by a Fourier series as a function with a period $T$. Customarily the time history is extended by a series of zeroes to give $n$ equal increments of $\Delta t$, thus deriving computational efficiency in using the fast Fourier transform (Newland, 1975). The frequency band is divided into corresponding intervals of $\Delta\omega = 2\pi/T$. Equations (3.38) and (3.39) can then be written in discrete form

$$P(\omega_l) = \Delta t \sum_{j=0}^{n-1} R(t_j) \exp\left(-2\pi i \frac{l_j}{n}\right) \qquad l = 0, 1, \ldots n - 1 \tag{3.40}$$

$$R(t_j) = \frac{\Delta\omega}{2\pi} \sum_{l=0}^{n-1} P(\omega_l) \exp\left(2\pi i \frac{l_j}{n}\right) \qquad j = 0, 1, \ldots n - 1 \tag{3.41}$$

where

$$t_j = j \, \Delta t = jT/n \tag{3.42}$$

$$\omega_l = l \, \Delta\omega = 2\pi l/T \tag{3.43}$$

The equation of motion can be expressed in the frequency domain as

$$\{-\omega_l^2[M] + i\omega_l[C] + [K]\}\{x(\omega_l)\} = \{P(\omega_l)\} \tag{3.44}$$

The values of frequency domain displacements $x(\omega_l)$ can then be expressed in the time domain using equation (3.41) rewritten as

$$\{x(t_j)\} = \frac{\Delta\omega}{2\pi} \sum_{l=0}^{n-1} x(\omega_l) \exp\left(2\pi i \frac{l_j}{n}\right) \tag{3.45}$$

This method enables frequency-dependent soil properties, damping and stiffness, to be used while retaining the benefits of the time domain solution.

## 3.11. Secondary structure response

Simple rules for mechanical equipment are given in Chapter 11. Where more accurate design information is required for secondary structure mounted on a building it is customary to produce 'floor response spectra' at the point or points of attachment. The floor response spectrum is simply the response of a single degree of freedom system mounted at this point and provides the design specification for the equipment.

The floor response spectrum is normally produced by exciting the main structure with a broad band ground motion record. The resulting displacement–time record at the point of attachment is then used to generate a response spectrum. Ideally this would be repeated several times so that the design response spectrum could envelope the resultant floor response spectra.

This process assumes that the secondary structure is much smaller than the main structure, so that it does not react on it significantly. It is conservative as long as the elastic properties of both the main and the secondary structures are known accurately. In practice errors in the estimation of structural frequencies may be quite large so that it is customary to allow for a margin of error in calculating both quantities. Unfortunately this frequently means that the possibility that the two structures' natural frequencies coincide has to be allowed for, resulting in high estimates of response for the secondary structure. This process may be compounded where there are tertiary structures so that gross overestimates may occur.

If the secondary structure is substantial it will modify the behaviour of the main structure, especially if its natural frequency happens to be close to one of the main structure's lower modal frequencies. Where this is the case it becomes necessary to model the combined structure.

## 3.12. Capacity design for reinforced concrete structures

Although the analytical tools that are available to structural engineers are powerful, it is not generally expedient to base the design of building structures purely on analysis. Normal design practice, recognised in virtually all seismic design codes, takes the following principal steps.

(a) Equivalent static lateral forces are derived, which take into account ground motion, importance, system dynamics, type of structural system, ductility, subsoil behaviour and the distribution of mass.

(b) Using the derived lateral forces a linear analysis is carried out, giving forces and displacements.

(c) Detailing of the structure is then carried out, firstly to provide strengths to resist the forces calculated in (b) and secondly to provide specified levels of member ductility.

(d) Also in the course of detailing certain general provisions are made to avoid the formation of hinges in columns and to ensure that, under yield conditions at the ends of a beam, shear failure will not occur.

The weakness in this approach is that it does not address the problem of non-linear behaviour directly, since it assumes that the strengths allocated on the basis of elastic analysis will be adequate once yielding has occurred. This is not necessarily true.

Capacity design philosophy addresses the problem directly and obliges the engineer to design the structure in such a way that hinges can only form in predetermined positions and sequence. It is a procedure whereby the design process in which strengths and ductilities are allocated and the analysis are interdependent. The importance of elastic analysis is de-emphasised, and the code lateral forces, combined with approximate elastic analysis, are used to allocate strength throughout the structure in a rational manner. The capacity design procedure stipulates the margin of strength that is necessary for elements to ensure that their behaviour remains elastic.

A general description of the approach is given by Paulay (1983). The reason for the name 'capacity design' is that, in the yielding condition, the strength developed in a weaker member is related to the capacity of the stronger member.

The simplest example of capacity design is in the design of beams. In order to establish the maximum likely moments at each end of a beam under earthquake loading it is necessary to allow for the overstrength of reinforcing steel above that specified and the increase in tensile strength during repeated cycles due to strain hardening. An overstrength factor $\phi_0$ which takes these influences into account can be applied to the calculated ultimate strength, based on specified yield strength and actual areas of reinforcement.

Once the maximum values of moments at each end of a beam have been established, the associated shear forces, including the effect of gravity load, can be calculated. When the beam is designed for this shear demand, the possibility of an undesirable, non-ductile shear failure is avoided. Fig. 3.14 illustrates the distribution of beam shear arrived at in this way.

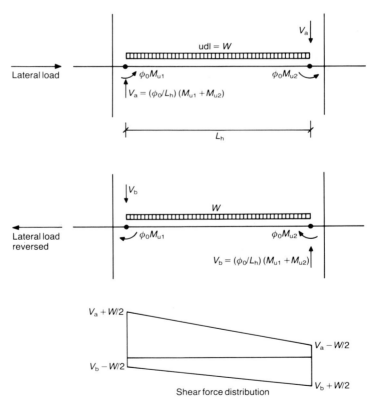

*Fig. 3.14. Capacity design beam shears*

It is important for designers to realise that the *strengthening* of the beam in bending can lead to *weakening* of the structure under earthquake forces by rendering it liable to shear failure, thereby dramatically reducing its ductility. Furthermore the acceptance of reinforcing steel of a higher yield strength than specified will have the same effect. It is also essential that all reinforcement in the beam and slab which can contribute to the ultimate strength is included in the calculation.

A typical value of $\phi_0$ may be taken as 1·4, except at roof level where a value of 1·1 may be used. Judgement should be used in assessing values as these values may be greatly exceeded where gravity loads are dominant.

In the case of columns the end moments $M_{u1}$ and $M_{u2}$ are derived from the ultimate moments of the connecting beams, calculated at the column centre line, and the elastic properties of the column. The overstrength factor $\phi_0$ is applied to these values. To allow for variations in the point of contraflexure due to the participation of higher modes of vibration in the response, a dynamic magnification factor $\omega$ is introduced so that the end moments for the column become

$$M_{\text{col}} = \omega \phi_0 M_u \qquad (3.46)$$

Figure 3.15 shows the way in which column moments are distributed.

The logical extension of capacity design principles to the detailing of columns would be to assume that there was no possibility of yielding whatever, so that confinement reinforcement at the ends could be omitted and splices could be located at floor level instead of at mid-

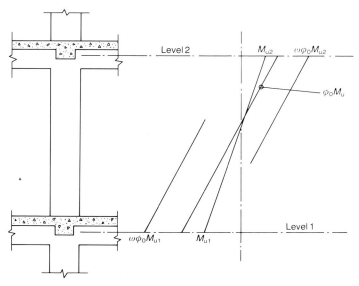

*Fig. 3.15. Capacity design bending moments for columns*

height (a potentially welcome move for steel fixers). However, this may not be permitted by the appropriate code.

Values of $\omega$ are given in NZS 3101:1982 as

$$\omega = 1\cdot3 < (0\cdot6T_1 + 0\cdot85) < 1\cdot8 \qquad (3.47)$$

where $T_1$ is the first-mode period of the structure in seconds for one-direction frames, and

$$\omega = 1\cdot5 < (0\cdot5T_1 + 1\cdot0) < 1\cdot9 \qquad (3.48)$$

for two-direction frames. These values are reduced at roof and ground floor levels to $1\cdot0$ for one-way frames and $1\cdot1$ for two-way frames.

For shear walls elastic analysis gives a poor representation of the moments that will be experienced once a hinge has formed at the base. For regularly shaped walls a moment envelope of the form shown in Fig. 3.16 is recommended for the distribution of moment reinforcement. For shear resistance the recommended shear provision at any level is taken as

$$V_{\text{wall}} = \omega_{\text{v}}\,\phi_0\,V_{\text{elastic}} \qquad (3.49)$$

where $\omega_{\text{v}}$ is a dynamic magnification factor and $\phi_0$ is an overstrength factor taking account of overstrength in the base hinge of the wall. $V_{\text{elastic}}$ will normally be based on code lateral forces. A value of $1\cdot4$ may be taken for $\phi_0$ and NZS 3101:1982 gives the values for $\omega_{\text{v}}$ given in Table 3.2.

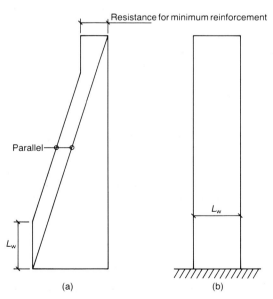

Fig. 3.16. *Capacity design bending moments for a reinforced concrete cantilever shear wall: (a) bending moment distribution; (b) wall elevation*

*Table 3.2. Dynamic magnification factors*

| Number of storeys | $\omega_v$ |
|---|---|
| 1–5 | $0.1N + 0.9$ |
| 6–9 | 1·5 |
| 10–14 | 1·7 |
| 15 and over | 1·8 |

Detailed guidance is given on the application of capacity design in NZS 3101:1982, and in simplified form in the Comité Euro-International du Breton (CEB) *Model code for seismic design of concrete structures*, April 1985. Although the capacity design approach probably represents the best practice in design for earthquakes, it may be more costly in terms of building construction cost than other simplified design code approaches.

## 3.13. Analytical procedures in practice

### 3.13.1. Selection of analytical method

Table 3.3 shows the principal design approaches used for buildings. The selection in any one case is a matter of practical economics and relates to the complexity and regularity of the structure, and whether floor response spectra for secondary structures are required.

### 3.13.2. Material and system values

Damping values may be taken from Table 3.4. The elastic dynamic modulus for concrete may be taken from Table 3.5. The dynamic modulus for structural steelwork may be taken as 200 kN/mm² $(29 \times 10^6 \text{ lbf/in}^2)$.

Where elastic analysis is used to study the behaviour of a reinforced concrete structure at a stress level where one or more members have cracked, ACI 318–83 (American Concrete Institute, 1983) recommends the use of an equivalent elastic stiffness $I_e$ for each cracked member given by

$$I_e = \left(\frac{M_{cr}}{M_a}\right)^3 I_g + \left[1 - \left(\frac{M_{cr}}{M_a}\right)^3\right] I_{cr} \tag{3.50}$$

where $I_g$ is the moment of inertia of the gross uncracked section, $I_{cr}$ is the moment of area of the cracked section transformed to concrete, $M_a$ is the maximum moment in a member at the stage when the deflection is being computed and $M_{cr}$ is the moment at first cracking, given by

$$M_{cr} = \frac{f_r I_g}{y_t} \tag{3.51}$$

with $y_t$ the distance from the centroidal axis of the cross-section to the extreme tension fibre and $f_r$ the modulus of rupture of the concrete.

63

*Table 3.3.   Approaches to the analysis of building response to ground motion*

| Input | Source | Linear or non-linear | Comment |
|---|---|---|---|
| Equivalent static forces | Design code | Linear | Static analysis, suitable for structures of regular form |
| Equivalent static forces | Capacity design code | Linear | Member design forces allocated on capacity design principles |
| Response spectrum | Design code Library | Linear | Economical, reasonably accurate method for regular or irregular structures |
| | | Non-linear | Economical, not very accurate |
| Time history (structure) | Library (recorded) Computer generated (artificial) | Linear | Step-by-step integration is expensive but will provide platform spectra for secondary structures Several analyses needed Expensive |
| | | Non-linear | As for linear but very expensive |
| Power spectrum | Library Kanai–Taijimi spectrum | Linear | Random vibration analysis, not used much at present, but simple and economical |
| | | Non-linear | Random vibration analysis using equivalent linearisation Only suitable for simple structures |
| Time history (soils) | Library (recorded) Computer generated (artificial) | Linear | Complex frequency solution Suitable for soil dynamics |

*Table 3.4.   Damping values*

| Structure type | Elastic stress level: % of critical | Inelastic stress level: % of critical |
|---|---|---|
| Reinforced concrete | 4 | 7 |
| Prestressed concrete | 2 | 5 |
| Steel-welded or friction grip bolts | 2 | 4 |
| Steel-bolted joints | 4 | 7 |

*Table 3.5. Dynamic modulus for concrete*

| Cube strength: N/mm² (lbf/in²) | $E$(dynamic): kN/mm² (lbf/in² × 10⁶) |
|---|---|
| 20 (2900) | 23 (3·3) |
| 25 (3630) | 25 (3·6) |
| 30 (4350) | 27 (3·9) |
| 35 (5080) | 28 (4·1) |
| 40 (5800) | 30 (4·4) |

## 3.14.  Modelling of building structures

Floors are normally considered to be rigid in plane so that lateral displacements of all connected members are geometrically related, with the floor having two translational and one horizontal torsional degrees of freedom.

Shear walls introduce additional degrees of freedom as the bending displacements in plane need to be accounted for, although these may be taken out of the analytical model by dynamic reduction.

Dynamic reduction (alternatively referred to as 'static condensation') is a process by which the number of degrees of freedom in a model can be reduced. The computing process, described by Bathe & Wilson (1976), successively modifies the equations of equilibrium to omit unwanted degrees of freedom. For example where a structure consists of a linked frame and a shear wall, it may be unnecessary to obtain the rotations of the shear wall at each floor level, so that they could be condensed out. The process should not be taken so far that the model is unable to represent the principal modes of response.

Dynamic reduction is an important weapon in the analyst's armoury and can achieve major cost savings in computing time for complex models.

In the modelling of X braces in structures where the diagonal members are insufficiently stiff to resist significant compressive forces, the diagonals can be modelled with one-half of their actual stiffness in tension.

For time history analysis, particularly when it is non-linear, the time step used may be critical. A simple criterion for determining that the step is sufficiently small is that repeating the analysis using the time step halved does not alter the results by more than 10%.

The torsional stiffness and centre of rigidity of a building structure may be calculated as follows

$$K_t = \sum_i (k_{yi} x_i^2 + k_{xi} y_i^2) - X^2 \sum_i k_{yi} - Y^2 \sum_i k_{xi} \qquad (3.52)$$

where $x_i$ and $y_i$ are the coordinates of the $i$th wall or column element, $k_{xi}$ and $k_{yi}$ are the lateral stiffnesses of the $i$th wall or column in the $X$ and $Y$ directions, $K_t$ is the torsional rigidity and $X$ and $Y$ are the coordinates of the centre of rigidity, given by

$$X = \frac{\sum_i x_i k_{yi}}{\sum_i k_{yi}} \qquad (3.53)$$

$$Y = \frac{\sum_i y_i k_{xi}}{\sum_i k_{xi}} \qquad (3.54)$$

### 3.15. Examples
#### 3.15.1. *Single degree of freedom, response spectrum*
Using the top storey of the structure in Table 3.6, with a fixed base

$$\text{mass} = 1.225 \times 10^6 \text{ kg}$$
$$\text{stiffness} = 3.24 \times 10^6 \text{ N/m}$$

then

$$\omega = \left(\frac{k}{m}\right)^{1/2} = \left(\frac{3.24 \times 10^8}{1.225 \times 10^6}\right)^{1/2} = 16.26 \text{ rad/s}$$

$$f = \frac{\omega}{2\pi} = \frac{16.26}{2\pi} = 2.59 \text{ Hz}$$

The response to a ground motion with a peak acceleration of 0.35$g$ is required, based on the design spectrum in Fig. 3.11. The displacement corresponding to 2.59 Hz is 6 cm. However, the spectrum is normalised to 0.5$g$ peak ground motion (asymptotic to this value at high frequencies) so that the response to 0.35$g$ is $(0.35 \times 6.0)/0.5 = 4.2$ cm.

#### 3.15.2. *Multiple degrees of freedom, response spectrum*
Using the model given in Table 3.6, and the design spectrum given in Fig. 3.10, from equations (3.16) and (3.18) the modal participation factors and modal displacements can be derived as

$$\gamma_1 = 2858$$
$$\gamma_2 = 1326$$
$$\gamma_3 = 900$$
$$\gamma_4 = 835$$

$$Y_1 = 2858 \times /S_d(4.66) = 2858 \times 1.6 = 4573 \text{ cm}$$
$$Y_2 = 1326 \times /S_d(10.92) = 1326 \times 0.23 = 305 \text{ cm}$$
$$Y_3 = 900 \times /S_d(18.99) = 900 \times 0.05 = 45 \text{ cm}$$
$$Y_4 = 835 \times /S_d(23.68) = 835 \times 0.03 = 25 \text{ cm}$$

The modal responses can now be tabulated using equation (3.19) in Table 3.7. Modal values for the fourth mode are negligible.

Table 3.6.  *Mass, stiffness and modal properties of a 10 storey building structure**

| Level | Mass: kg | Stiffness: N/m | Normalised mode shapes | | | |
|---|---|---|---|---|---|---|
| | | | Mode 1 | Mode 2 | Mode 3 | Mode 4 |
| 1 | $1 \cdot 225 \times 10^6$ | $2 \cdot 5 \times 10^9$ | $0 \cdot 2485 \times 10^{-4}$ | $0 \cdot 6857 \times 10^{-4}$ | $-0 \cdot 1187 \times 10^{-3}$ | $-0 \cdot 1874 \times 10^{-3}$ |
| 2 | $1 \cdot 225 \times 10^6$ | $2 \cdot 5 \times 10^9$ | $0 \cdot 4943 \times 10^{-4}$ | $0 \cdot 1331 \times 10^{-3}$ | $-0 \cdot 2164 \times 10^{-3}$ | $-0 \cdot 3234 \times 10^{-3}$ |
| 3 | $1 \cdot 225 \times 10^6$ | $2 \cdot 5 \times 10^9$ | $0 \cdot 7349 \times 10^{-4}$ | $0 \cdot 1900 \times 10^{-3}$ | $-0 \cdot 2758 \times 10^{-3}$ | $-0 \cdot 3704 \times 10^{-3}$ |
| 4 | $1 \cdot 225 \times 10^6$ | $1 \cdot 024 \times 10^9$ | $0 \cdot 1303 \times 10^{-3}$ | $0 \cdot 3015 \times 10^{-3}$ | $-0 \cdot 3019 \times 10^{-3}$ | $-0 \cdot 2368 \times 10^{-3}$ |
| 5 | $1 \cdot 225 \times 10^6$ | $1 \cdot 024 \times 10^9$ | $0 \cdot 1837 \times 10^{-3}$ | $0 \cdot 3700 \times 10^{-3}$ | $-0 \cdot 1977 \times 10^{-3}$ | $0 \cdot 5556 \times 10^{-4}$ |
| 6 | $1 \cdot 225 \times 10^6$ | $1 \cdot 024 \times 10^9$ | $0 \cdot 2324 \times 10^{-3}$ | $0 \cdot 3858 \times 10^{-3}$ | $-0 \cdot 8203 \times 10^{-5}$ | $0 \cdot 3107 \times 10^{-3}$ |
| 7 | $1 \cdot 225 \times 10^6$ | $1 \cdot 024 \times 10^9$ | $0 \cdot 2750 \times 10^{-3}$ | $0 \cdot 3465 \times 10^{-3}$ | $0 \cdot 1849 \times 10^{-3}$ | $0 \cdot 3574 \times 10^{-3}$ |
| 8 | $1 \cdot 225 \times 10^6$ | $3 \cdot 24 \times 10^9$ | $0 \cdot 3871 \times 10^{-3}$ | $0 \cdot 6625 \times 10^{-4}$ | $0 \cdot 5249 \times 10^{-3}$ | $-0 \cdot 2527 \times 10^{-3}$ |
| 9 | $1 \cdot 225 \times 10^6$ | $3 \cdot 24 \times 10^8$ | $0 \cdot 4673 \times 10^{-3}$ | $-0 \cdot 2439 \times 10^{-3}$ | $0 \cdot 1604 \times 10^{-3}$ | $-0 \cdot 3271 \times 10^{-3}$ |
| 10 | $1 \cdot 225 \times 10^6$ | $3 \cdot 24 \times 10^8$ | $0 \cdot 5092 \times 10^{-3}$ | $-0 \cdot 4441 \times 10^{-3}$ | $-0 \cdot 4409 \times 10^{-3}$ | $0 \cdot 2920 \times 10^{-3}$ |
| Natural frequency: rad/s | | | 4·662 | 10·92 | 18·99 | 23·68 |

* Single translational degree of freedom at each level; storey height, 3·0 m; mode shapes normalised according to equation (3.17).

67

*Table 3.7. Modal responses for a 10 storey building structure*

| Level | $x_1$: cm | $x_2$: cm | $x_3$: cm | $x^*$: cm |
|-------|-----------|-----------|-----------|-----------|
| 1 | 0·11 | 0·02 | 0·0 | 0·11 |
| 2 | 0·23 | 0·04 | −0·01 | 0·23 |
| 3 | 0·34 | 0·06 | −0·01 | 0·35 |
| 4 | 0·60 | 0·09 | −0·01 | 0·61 |
| 5 | 0·84 | 0·11 | −0·01 | 0·85 |
| 6 | 1·06 | 0·12 | 0·0 | 1·07 |
| 7 | 1·26 | 0·11 | 0·01 | 1·27 |
| 8 | 1·77 | 0·02 | 0·02 | 1·77 |
| 9 | 2·14 | −0·07 | 0·01 | 2·14 |
| 10 | 2·33 | −0·14 | −0·02 | 2·33 |

\* Square root of the sum of the squares.

## 3.16. Bibliography

Bathe, K.-J. & Wilson, E. L. (1976). *Numerical methods in finite element analysis*. Englewood Cliffs: Prentice-Hall.

Chopra, A. K. (1982). *Dynamics of structures: a primer*. Berkeley: Earthquake Engineering Research Institute.

Clough, R. W. & Penzien, J. (1975). *Dynamics of structures*. New York: McGraw Hill.

Newland, D. E. (1984). *An introduction to random vibrations and spectral analysis*. London: Longman.

Newmark, N. M. & Hall, W. J. (1982). *Earthquake spectra and design*. Berkeley: Earthquake Engineering Research Institute.

# Chapter 4

# Isolation and energy absorbers

'Je plie et ne romps pas' (I bend and
break not). *Le chêne et le roseau*,
Jean de la Fontaine, 17th century

## 4.1. Introduction

For many years engineers have sought alternative ways to achieve
earthquake resistance. Many of these have been impractical—
suspending buildings on chains, seating them on rocking blocks, suspending large pendulums from the roof, digging enormous trenches
around the foundations or using high velocity gas jets to create reac-
tive forces. The most practicable methods involve three alternative
approaches: isolation, energy-absorbing dampers and, perhaps, active
control.

Active control utilises the feedback from sensors measuring the
response of a structure to control the behaviour of structural elements through mechanical actuators. Although this method has been
used in practice to control wind-induced vibrations of tall structures
it has not been applied to earthquake-resistant structures (although
there are numbers of theoretical studies). For this reason it is not
dealt with here. It is difficult to predict whether or not future
improvements in control technology and reliability will bring active
control into common use.

Energy absorbers and isolation have been used in practice both
separately and in conjunction, although none has yet been subjected
to the most searching examination of exposure to strong ground
motion.

The essential concepts of either technique are attractive. Isolation
attempts to cut off, or to limit, transmission of earthquake forces to
the structure. Energy absorbers take over the role of energy-
absorbing portions of the structure, so that after a major earthquake
the structure is undamaged and the absorbers, which are not involved
in supporting the structure, may be replaced if necessary. Each technique involves a different approach, but in practice they are comple-
mentary.

The use of isolation or energy absorption on a building is rarely
justified by a substantial saving in first cost. Justification is likely to
come from increased levels of safety, lower levels of structural
damage in extreme events or from lower damage levels for building
contents. In the nuclear industry isolation has been used to achieve

69

standard designs for the superstructure in areas of differing seismicity.

## 4.2. Isolation

The pure isolation method uses some form of soft structural support such as springs so that earthquake forces are only partially transmitted to the building (Fig. 4.1). The perfect example of isolation is a ship which avoids all horizontal components of ground motion as the water on which it rests cannot transmit them. Isolation in practice is limited to a consideration of horizontal forces, to which buildings are most sensitive. Vertical isolation is less needed and much more difficult to implement.

Figure 4.2 shows basic arrangements for providing sprung supports to a building structure. Arrangement 4.2(a) is the more expensive arrangement but may be preferred for architectural reasons. The alternative 4.2(b) is particularly suitable where there is a basement. Fig. 4.3 shows the types of laminated rubber spring used for this application. These springs have been used in a number of buildings and have high vertical stiffness. The lead plug introduced into the spring shown in Fig. 4.3(a) is an energy-absorbing damping mechanism.

Figure 4.4 shows the influence of introducing springs on the acceleration response spectra for two earthquakes. It is clear that the acceleration forces on the building will be considerably reduced for the El Centro event (a relatively common type of spectrum), but equally clear that for the Bucharest earthquake (which had an unusual frequency content) matters have been made much worse. Generally a reasonable estimate can be made of the spectral properties of future ground shaking on a particular site, but for the application of isolation a conservative view should be taken. Thus isolation can show substantial reductions in response, but depends for its effectiveness on knowing in advance the kind of frequency content that the earthquake will have.

An alternative type of isolation device is the spring combined with a sliding support, which has been used on nuclear reactors. The effect is to cut off input accelerations, as a fraction of $g$, at a value approximately equal to the coefficient of friction of the sliding surface. Materials are available with coefficients of friction in the range 0·03–0·05 so that this represents a substantial reduction from strong motion inputs.

Practical constraints limit the range of spring or sliding support that can be used. For springs it is desirable to limit the relative displacement between the structure and the ground to avoid fracture. Similarly with sliding surfaces there is a possibility of falling off the support or finishing up with an inconveniently large residual displacement. Problems may also arise in linking service pipes across the moving joint and with detailing lift wells.

Provisions for wind, or other lateral forces, can also present difficulties, as the soft springs may deflect excessively. Ideally the base of

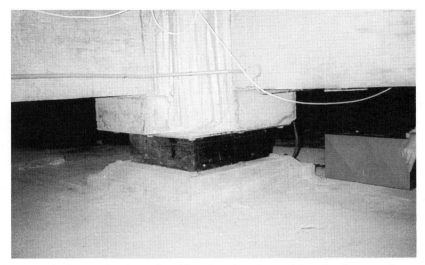

*Fig. 4.1. Isolating laminated rubber bearing in position in the crawl space under the William Clayton Building, Wellington, New Zealand*

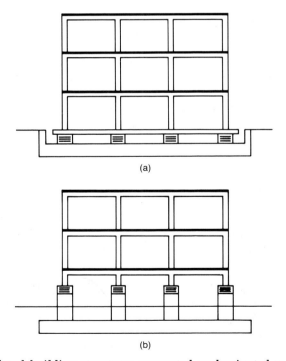

(a)

(b)

*Fig. 4.2. Isolated building structures supported on laminated rubber springs: (a) double foundation with crawl space; (b) springs at mid-height of the lower storey columns*

71

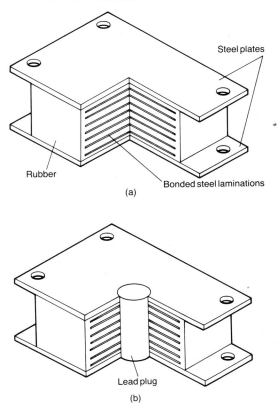

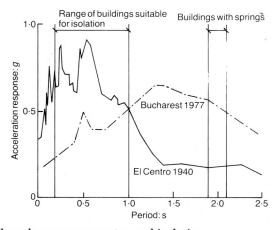

*Fig. 4.3. Types of laminated rubber springs used in isolation: (a) undamped; (b) damped*

*Fig. 4.4. Earthquake response spectra and isolation*

the structure would be rigidly fixed under all conditions except when an earthquake occurs, and one proposed solution to this is to have breakable links, or 'mechanical fuses', which fracture when subjected to moderate earthquake forces. This induces some undesirable frequency content into the structure at the time of fracture which might adversely affect the building contents.

## 4.3.   Energy absorbers

In order to function absorbers have to link points with relative displacement in response to ground motion. Some structural configurations in which this can be achieved are shown in Fig. 4.5 and some types of absorber in Figs 4.6 and 4.7. The arrangements shown in Figs 4.5(a) and 4.5(b) are forms of pure energy absorption without involving any form of isolation or other form of introducing springs into the structure. Fig. 4.8 shows the response of a 10 storey structure incorporating energy-absorbing dampers in the arrangement shown in Fig. 4.5(b), the responses shown being those from five different ground motions of similar frequency content and peak acceleration. Response ratios are calculated as a fraction of displacement response with the shear wall and frame rigidly connected.

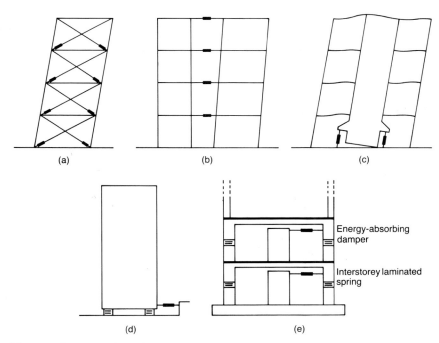

*Fig. 4.5. Structural configurations with energy-absorbing dampers: (a) braced frame; (b) stiff core and frame: distributed energy-absorbing dampers; (c) rocking stiff core and frame; (d) isolation with energy-absorbing damping; (e) two-level filter system*

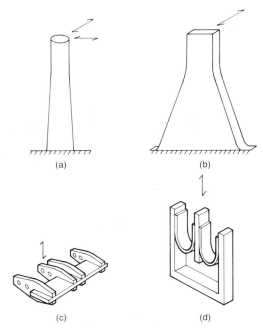

(a)                    (b)

(c)                    (d)

*Fig. 4.6. Yielding steel energy-absorbing dampers: (a) tapered circular section;*
*(b) tapered flat plate; (c) torsional system; (d) U-strip bending*

*Fig. 4.7. Steel energy-absorbing damper in place on Union House, Auckland,*
*New Zealand*

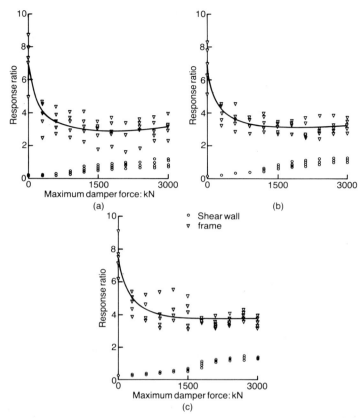

*Fig. 4.8. Earthquake response using energy-absorbing dampers linking the stiff shear wall system to the framed structure*

It can be seen that at around a damper force of 1500 kN the frame response has been reduced by about 50% from the free-standing condition (zero damper force). For the rigidly connected frame and shear wall condition effectively all the base shear is taken by the shear wall, so that all values of shear wall response below unity represent reductions in response. Clearly, as the damper force is increased, a range of options exists on the distribution of forces between shear wall and frame. This is an extremely valuable design option because, in practice, stability considerations on the shear wall are frequently critical.

Energy absorbers deal with the problem of response to wind forces without difficulty. They can be simply designed to remain elastic up to the maximum wind force and only to yield above this level.

## 4.4.   Combined isolation and energy-absorbing damping

If the combined system, illustrated in Fig. 4.5(d), is studied with varied parameters, a response plot similar to that shown in Fig. 4.9 is

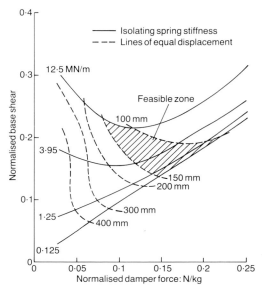

*Fig. 4.9. Isolation combined with energy-absorbing damping: mean response ratios for a 10 storey reinforced concrete building to five simulated earthquakes with peak accelerations of 0·3g (structural damping, 0·05)*

obtained. This gives the response of a 10 storey structure, the responses shown being the mean of five different ground motions, each of similar frequency content and peak acceleration. Damper force is normalised to the mass of the building, and base shear is normalised to the value for the structure rigidly connected to the ground, so that its value represents the effectiveness of the system.

The balance to be struck between the design parameters of spring value and damping on the one hand and response parameters base shear and displacement on the other is clear from Fig. 4.9. The feasible zone is shown as lying arbitrarily between displacement values of 100 mm and 150 mm but in practice considerations of suitable load-carrying springs and the economics of providing damping will provide further constraints.

The important point shown by such response studies is that the performance of an isolation–damper system is as dependent on choosing the value of damping as it is on the spring value. Further, as damping is increased, there is a general tendency to be less sensitive to the spectral content of the earthquake.

## 4.5. Design of damper–isolation systems

Current practice concentrates on modifying the first period of the building to approximately 2 s, although this should be evaluated against the spectral properties of the design earthquake.

Analysis can be carried out firstly by non-linear time history with the building and support system modelled as a single degree of

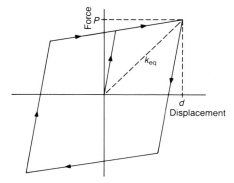

*Fig. 4.10. Equivalent stiffness for a hysteretic system*

freedom system. Once a support system has been designed to give the best response as a single degree of freedom system, then the full system can be modelled and reanalysed by time history. The effect of the damper can either be modelled by using equivalent viscous damping, or by non-linear time history analysis. An alternative approach is to generate inelastic design response spectra for single degree of freedom oscillators which possess the hysteretic response properties of the bearing and damper to be used. This procedure is described in detail by Mayes, Jones, Kelly & Button (1984).

Either of the foregoing approaches is based on elastic response for the isolated/damped structure. In general it is the objective of such a design to ensure that the structure remains in the elastic range, or that any excursions into the non-linear range of response are small.

An approximate method of analysis can be used for preliminary design, treating the system as having a single degree of freedom. This uses equivalent stiffness and damping for a linear system which approximately reproduce the peak response of a non-linear system. Fig. 4.10 shows the simple approach to equivalent stiffness, $k_{eq}$, which is taken as $P/d$.

Equivalent viscous damping is then derived by equating the energy absorbed by hysteresis with the energy absorbed by equivalent viscous damping, which gives

$$\zeta_{eq} = \frac{\text{area of hysteretic loop}}{2\pi k_{eq} d^2} \qquad (4.1)$$

Where the spring system incorporates viscous damping already this is added to the equivalent viscous damping value calculated from equation (4.1).

The response of the system to ground motion can then be derived from the response spectrum, an iteration process for displacement being required to arrive at a compatible solution. This method is one of two examined by Karamchandani & Reed (1986) and is shown to give a reasonable approximation to the response calculated from time history analysis.

## 4.6.  Application of isolation and energy-absorbing damping

Two guides on the use of the technique exist: Blakely, Charleson, Hitchcock, Megget, Priestley, Sharpe & Skinner (1979) and Mayes *et al.* (1984). These describe the general principles and practical application. Blakely *et al.* (1979) recommend the provision of some ductility and strength in the main structure in excess of that demanded by the analysis.

Laminated rubber bearings are dealt with in British Standard BS 6177:1982, and tables of the available standard units are obtainable from the manufacturers. The current maximum vertical load for a single bearing is approximately 15 000 kN (1500 t), but there is no limit on capacity other than the availability of sufficiently large moulds. Some manufacturers provide lead-cored bearings but these are patented and not available from all bearing suppliers. High damping rubber for bearings has been developed and used in practice (Derham, 1986).

Standard types of hysteretic steel dampers are not available commercially at the time of writing. They can be manufactured by non-specialist firms without difficulty, however, and guidance on their design can be found from Tyler (1978).

*Fig. 4.11. Union House, Auckland, New Zealand, uses the lateral flexibility of sleeved piles to provide lateral isolation: forces in the superstructure are resisted by the external diagonal bracing*

## 4.7. Bibliography

Applied Technology Council. *Proc. Seminar and Workshop on Base Isolation and Passive Energy Dissipation, San Francisco, March 1986.* Redwood City: A.T.C.

Blakely, R. W. G. *et al.* (1979). Recommendations for the design and construction of base isolated structures. *Bull. N.Z. Soc. Earthquake Engng* 12, June, No. 2, 136–157.

Mayes, R. L. *et al.* (1984). Design guidelines for base isolated buildings with energy dissipators. *Earthquake Spect.* 1, Nov., No. 1, 41–74.

# Chapter 5

# Conceptual design

'Some of the major problems relating
to earthquake design are created by
the original design concept chosen by
the architect.' Henry Degenkolb

The scope this chapter covers

(a) the anatomy of a building
(b) overall form
(c) framing systems
(d) costs.

## 5.1. Design objectives

Nothing within the power of a structural engineer can make a badly conceived building into a good earthquake-resistant structure. Decisions made at the conceptual stage are difficult to modify so that it is essential that their full consequences are understood in terms of performance and costs as early as possible.

## 5.2. Anatomy of a building

The functioning parts of a building affect the way in which it can accommodate its structural skeleton. For this reason it is useful to consider the principal division of functions and how they affect the structure.

The principal categories of building use can be considered in a vertical direction as given in Table 5.1. The vertical divisions of the building principally provide problems, making it difficult to avoid irregularities in mass or stiffness. However, the service cores and exterior cladding provide an opportunity to incorporate shear walls or braced panels. One of the main objectives in early planning is to establish the optimum locations for service cores and for stiff structural elements that will be continuous to the foundation.

It is not unusual to find that structural and architectural requirements are in conflict at the concept planning stage but it is essential that a satisfactory compromise is reached at this time.

## 5.3. Overall form

The desirable aspects of building form are simplicity, regularity and symmetry in both plan and elevation. These properties all contribute to a more even and more predictable distribution of earth-

81

*Table 5.1. Categories of building use*

| | |
|---|---|
| Basement | Car parking, storage, mechanical and electrical plant |
| Street level | May be used quite differently from the rest of the building, commonly leading to a greater than typical storey height and a need for unobstructed floor space, e.g. in hotels the street level may be used for reception, conference and restaurant areas in contrast with the regular pattern of rooms on the typical floors; in office buildings the street level may include shops, banks, restaurants etc. |
| Typical floors | Repetitive standard levels |
| Roof structures | Mechanical and electrical plant, lift motor room, water tanks etc. |
| *In plan* Service cores | Stairs, lifts, toilets and pipe ducts, which are frequently grouped together |
| Usable floor | Clear spaces, usually modular. |
| Exterior cladding | Provides opportunities for bracing, shear walls. |

quake forces in the structural system. Any irregularity in the distribution of stiffness or mass is likely to lead to an increased dynamic response.

Torsional forces from ground motion are not commonly of great concern unless the building has an inherently low torsional strength. However, torsion also arises from eccentricity in the building layout. The effective force exerted by lateral ground movement acts at the centre of gravity of each floor creating a torsional moment about the centre of structural resistance, and this will have to be dealt with in addition to the torsional component of ground motion. This is illustrated in Table 5.2.

Buildings which are tall in relation to their base width will generate high forces at the base due to the overturning moment. Buildings with height-to-width ratios of about 4 are common, whereas those with a ratio of 6 are rare. It is probably within the range 4–6 that the problems arising from overturning forces become critical. The high forces may lead to foundation uplift or to unduly high tensile or compressive forces in columns.

Buildings of great length or plan area may not respond in the way calculated. Analysis customarily assumes that the ground moves as a rigid mass over the base of the building, but this is only a reasonable assumption for a small area. The ground is elastic and the propagation of seismic waves is not instantaneous. If different parts of the building are being shaken out of step with each other, additional, incalculable, stresses are being imposed, and this effect increases with size.

*Table 5.2. Plan configuration*

| Plan shape* | For | Against |
|---|---|---|
| | Symmetric overall form | Low torsional modulus<br>High eccentricity |
| | Symmetric overall form<br>No eccentricity<br>Adequate torsional modulus | |
| | Symmetric overall form<br>No eccentricity<br>High torsional modulus | |
| | Symmetric overall form | Low torsional modulus<br>High eccentricity |
| | Symmetric overall form<br>No eccentricity<br>High torsional modulus† | |
| | No eccentricity | Very low torsional modulus<br>Asymmetric overall form |
| | No eccentricity | Very low torsional modulus<br>Asymmetric overall form |

* M, centre of mass; R, centre of resistance.
† Core and shear wall stiffnesses are balanced in the $X$ direction.

Some guidance on maximum plan dimensions for seismic regions is given in the Russian earthquake design code Gosstroi, USSR (1969). Large panel buildings without frames are limited to 60 m (197 ft). Buildings with good quality masonry, or masonry with reinforced concrete walls, may not exceed 80 m (262 ft) in region 8 or 60 m in region 9. These values are reduced for masonry of lower quality. The regions are defined by the expected maximum earthquake intensity, modified in some cases by the importance of the building.

The effect of out-of-step vibration also occurs in any building founded on subsoil where there is a marked discontinuity. As an extreme example, a structure founded partly on rock and partly on alluvium would be severely stressed at the interface between the two, each material tending to vibrate differently.

The solution to many of the problems arising from buildings of irregular form is to divide them into regular shapes by means of joints. Such joints are required to be sufficiently wide to avoid damage by impact during earthquakes.

Buildings on sloping sites tend to pose torsional problems due to the varying column stiffnesses; this is illustrated in Fig. 5.1. The solution to this is to provide additional stiffening elements at the low end of the site to bring the centre of resistance as close as possible to the centre of mass.

Proposals have been made in favour of reducing the stiffness of the lower storey, the so-called 'soft storey' approach (Fintel & Khan, 1969). The reasoning behind this is that by so doing a reduced dynamic force is transmitted to the superstructure. However, this argument is based on simple elastic analysis and when realistic inelastic and geometrical non-linear effects are taken into account this approach is shown to be potentially disastrous (Chopra, Clough & Clough, 1973). Arnold (1984) gives a discussion of the reasons why architectural requirements lead towards the soft storey type of structure and the problems inherent in the architect–engineer conflict of interest.

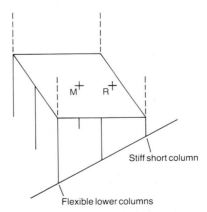

*Fig. 5.1. Sloping site (M, centre of mass; R, centre of resistance)*

## 5.4. Framing systems

Considerations of the overall concept and of the detailed framing are not independent, and at the planning stage some consideration will need to be given to the framing layout.

Basic types of shear wall systems are shown in Fig. 5.2. The bidirectional 'egg crate' system is suitable for tall buildings but unsuited to office buildings which need large unobstructed areas. The structural core and frame can be used for buildings up to about 40 storeys and above this height the single-framed tube should be used, with the 'tube-in-tube' system being used for the highest buildings.

The shear walls shown in Fig. 5.2 may be either pierced or unpierced reinforced concrete, or diagonally braced frames. The essential feature is that they are structural elements that are much stiffer than the adjoining framed structure. Within the basic categories of frame–shear wall systems, many hybrid systems can be produced to suit the particular needs of a project.

A critical feature in relating the framed structure to the shear wall system is the distribution of vertical load. Because nearly all the lateral seismic force is resisted by the shear walls while the structure is behaving elastically, and a lesser but still substantial portion after yielding, the stability of shear walls requires a high vertical load.

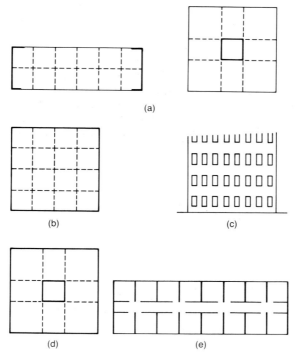

(a)

(b)

(c)

(d)

(e)

*Fig. 5.2. Shear wall and frame systems: (a) frame and shear wall; (b) tube; (c) elevation of the tube structure; (d) tube-in-tube structure; (e) egg crate*

*Fig. 5.3. This building in Christchurch, New Zealand, uses projecting beam stubs to improve the performance of external beam–column joints*

*Fig. 5.4. This Los Angeles steel structure uses closely spaced external framing to form a load-resisting system in the exterior surfaces of the building*

*Fig. 5.5. The Alcan Building in San Francisco uses the external surface to incorporate a diagonal bracing system which forms an architectural feature*

There is usually no great problem in achieving sufficient strength in the shear wall itself; the difficulty arises at foundation level where there might be uplift at the footing level.

In designing reinforced concrete shear walls an understanding of the ductility requirement is necessary, some plan shapes being more effective than others. The two principal approaches to ductility are illustrated in Fig. 5.6 and examined in more detail in Chapter 7. For braced steel frames greater ductility can be achieved with an eccentric connection design which is described in Chapter 8.

In planning the framed structure the relationship between members at beam–column junctions becomes critical. Fig. 5.7 shows possible modes of failure under lateral load and it is clear from this that yielding in the major earthquake must occur in the beams and not in the columns. Considering a single beam–column connection such as that in Fig. 5.8 it follows that

$$M_{b1} + M_{b2} = M_{c1} + M_{c2} \tag{5.1}$$

The problem posed by equation (5.1) increases as beam spans increase leading to a need for greater continuity reinforcement at the support, and consequently a greater ultimate moment. For grid or flat plate floors it becomes extremely difficult to establish how and where yield in the grid beams or slab takes place and at what level of force it will occur. Another case also posing difficulty is the spandrel beam

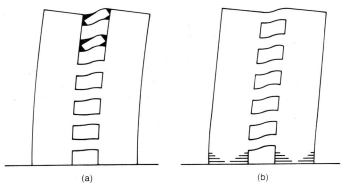

Fig. 5.6. Ductility in shear walls: (a) in coupled shear walls, overall ductility can be activated by yielding in the coupling beams; (b) ductility by yielding at the base

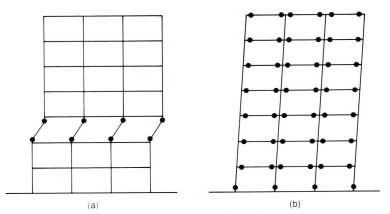

Fig. 5.7. Frame failure modes: (a) local failure by columns yielding before beams; (b) overall failure by beams yielding before columns

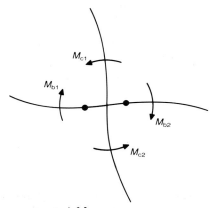

Fig. 5.8. Bending moments at yield

88

detail illustrated in Fig. 5.9. Here the beam is oversized for architectural reasons and may have an unnecessarily high yield moment.

Floor slabs function as diaphragms in transferring lateral forces. Fig. 5.10 shows two possible floor plans. In the first case there is very little diaphragm action but in the second case it is clearly significant and the transfer of shear at each end wall may require additional reinforcement. The functioning of the slab as a diaphragm, or hori-

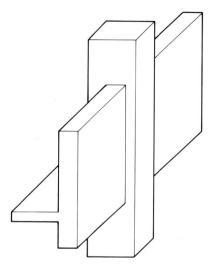

*Fig. 5.9. Spandrel beam*

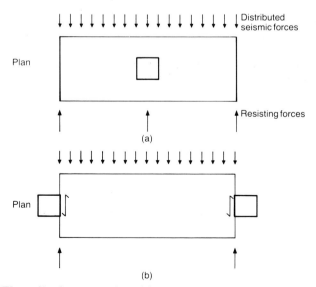

*Fig. 5.10. Floor diaphragm action: (a) case 1, relatively low diaphragm stress; (b) case 2, substantial diaphragm stress and high end shears*

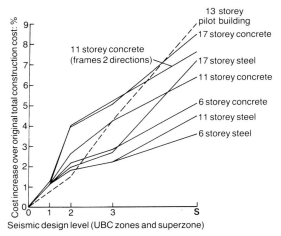

*Fig. 5.11. Seismic cost increase (after Whitman & Liao (1974)) (Uniform building code levels are defined as follows: 1, intensity MM5–6—minor damage; 2, intensity MM7—moderate damage; 3, intensity MM8+—major damage; 4, zone 3 and close to a major active fault)*

zontal beam, needs to be considered in the selection of floor type. Some fully or partially prefabricated systems have very little strength in horizontal shear or bending.

## 5.5. Cost of seismic design provision

Figure 5.11 shows some comparative cost estimates for the increase in construction cost for seismic design provision in Boston, USA. Obviously the costs will vary according to the baseline design provision for wind—if high wind forces are experienced then the percentage increase due to seismic design would be less.

Other estimates (Leslie & Biggs, 1972; Whitman, Biggs, Brennan, Cornell, de Neufville & Vanmarcke, 1975; Ferrito, 1985) generally lie within the upper and lower bound of Fig. 5.11 and agree that the proportional increase is less for structural steel, compared with reinforced concrete, and that for taller buildings the increase is greater.

## 5.6. Bibliography

Arnold, C. & Reitherman, R. (1982). *Building configuration and seismic design*. New York: Wiley.

# Chapter 6

# Design codes and lateral force design

'The purpose of this code is to provide minimum
standards to safeguard life or limb, health,
property, and public welfare . . .'.
*Uniform building code* (1982)

The scope of this chapter covers

(*a*) the development and philosophy of codes
(*b*) lateral forces
(*c*) ductility
(*d*) miscellaneous code requirements.

## 6.1. Design objectives

In most actively seismic areas, building construction is subject to a legally enforceable code which establishes minimum requirements. Even where this is not so, common practice or contractual requirements will require compliance with a code. In consequence it will form part of the normal design process that a set of minimum performance criteria will have to be met.

It is assumed in this chapter that the reader has identified a specific code. References to sources of codes are given at the end of the chapter.

## 6.2. Development of codes

Early codes were based directly on the practical lessons learned from earthquakes, relating primarily to types of construction. In some cases they placed limitations on the height of buildings. It was recognised that timber structures performed well, even the relatively tall Japanese pagodas, whereas plain masonry buildings performed poorly.

In 1909 following the Messina earthquake, which caused 160 000 deaths, an Italian commission recommended the use of lateral forces equal to 1/12th of the weight supported. This was later increased to one-eighth for the ground storey. The concept of lateral forces also became accepted in Japan although there was a division of opinion on the merits of rigidity as opposed to flexibility. After the 1923 Tokyo earthquake a lateral force factor of 1/10 was recommended and a 33 m height limit imposed. In California lateral force requirements were not adopted by statute until after the 1933 Long Beach earthquake.

The use of lateral forces in design became widely used, the value of the coefficients being based almost entirely on experience of earthquake damage.

In 1943 the City of Los Angeles related lateral forces to the principal vibration period of the building and varied the coefficient through the height of multi-storey buildings. By 1948 information on strong motion and its frequency distribution was available and the Structural Engineers Association of California recommended the use of a base shear related to the fundamental period of the building. The logic of this is demonstrated in Fig. 6.1. Once information was available on the spectra of earthquake ground motion, the arguments over flexibility versus rigidity could be resolved. The flexible structure was subjected to lower dynamic forces but was usually weaker and suffered larger displacements.

The next important step grew out of advances in the study of the dynamic response of structures. This led to the base shear being distributed through the height of the building according to its dynamic properties, and the formula used in the 1960 Structural Engineers Association of California code is still widely used (see equation (6.7)). At this stage lateral forces had undergone a quiet revolution from an arbitrary set of forces based on earthquake damage studies to a set of forces which, applied as static loads, will reproduce the peak dynamic response of the structure to the design earthquake.

This, however, is not quite the end of the matter for lateral loads, because structural response to strong earthquakes involves yielding of the structure so that the response is inelastic.

The effect of yielding in a structure under earthquake loading is twofold. On the one hand stiffness is reduced so that displacements

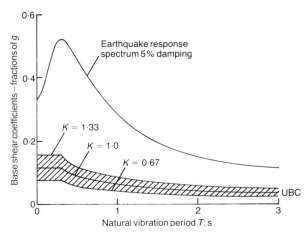

*Fig. 6.1. Comparison of base shear coefficients from an elastic response spectrum with the Uniform building code*

tend to increase, and on the other hand hysteretic yielding absorbs energy from the structure, increasing damping and reducing displacements. The two effects are roughly equal so that yielding does not have a large effect on displacement.

Figure 6.2 shows the relationship between lateral design forces for an elastic structure and for a yielding ductile structure. Much larger design forces are required for an elastic structure without ductility. Because it is found in practice that the increased cost of elastic design requirements is unacceptably large it is almost universally accepted that ductile design should apply for major earthquakes. Exceptions to this are made for structures of special importance, or where the consequences of damage are unacceptable.

Although modern codes contain much useful guidance on other matters, it is the calculation of lateral design forces and the means of providing sufficient ductility that constitute, in practice, the two most vital elements for the structural engineer.

## 6.3. Philosophy of design

The following philosophy of design is widely accepted in national and state building codes. Structures should, in general, be able to

- (a) resist minor earthquakes without damage
- (b) resist moderate earthquakes without structural damage, but with some non-structural damage
- (c) resist major earthquakes, of severity equal to the strongest that could be experienced in the area, without collapse but with some structural as well as non-structural damage.

Why should earthquake design permit substantial damage when it is not acceptable for other environmental loadings? The fundamental reason lies in the costs of seismic design provisions which would be

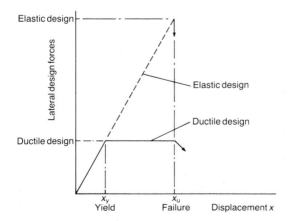

*Fig. 6.2. Lateral design forces and ductility (ductility factor, $x/x_y$; ductile capacity, $x_u/x_y$)*

excessive if the maximum design earthquake were to be resisted without damage. The acceptance of survival as the aim in a major earthquake means that the design objective becomes that of preserving the lives of the buildings' occupants.

## 6.4. Calculation of lateral forces

### 6.4.1. Principal factors

The factors taken into account in assessing lateral design forces are as follows. Although international standards vary in their application, these factors are common to most modern seismic codes

(a) Z, the zoning factor for regional seismicity
(b) I, the importance factor
(c) S, the subsoil factor
(d) K, the structural type factor
(e) T, the natural period of vibration of the building
(f) W, the applicable building mass.

### 6.4.2. Zoning factor Z

Seismic zoning assesses the maximum severity of shaking that is anticipated in the region. The steps between regions are usually fairly large: for example the *Uniform building code* (International Conference of Building Officials, 1982), USA, uses Z values of 1·0, 0·75, 0·375 and 0·188, and the People's Republic of China's code uses 0·9, 0·45 and 0·23. Normally the zoning will be laid down by a code, but outside the area of applicability the zoning status will need to be based on an assessment of the seismic hazard.

### 6.4.3. Importance factor I

It is customary to recognise that certain categories of building use should be designed for greater levels of safety, and this is achieved by specifying higher lateral design forces. Such categories are

(a) buildings which are essential after an earthquake—hospitals, fire stations, power-stations etc.
(b) places of assembly—schools, theatres etc.
(c) structures whose collapse would endanger the population— nuclear plant, dangerous chemical storage vessels, large dams etc.

Typically the value of *I* varies from 1·0 to 1·5 but values 2·0 and 4·0 are used in the USSR. Structures in category (c) are normally designed on a different basis, concentrating on reducing the risk of serious accident to an acceptable level.

### 6.4.4. Subsoil factor S

The effect of soft subsoil may be both to magnify ground motion and to lengthen its characteristic period of motion. The soil factor takes account of both the magnification and the interaction between building response and soil response. If the natural period of vibration of the building and the soil are close, resonance will occur.

The soil factor is typically in the range 1·5–2·0 for soft soils compared with a value of 1·0 for rock. An interesting development is that the Structural Engineers Association of California (1985) omits any consideration of resonance between building and soil from its recommended $S$ values and proposes three values

(a) $S = 1·0$, for rock-like material having a shear wave velocity greater than 2500 ft/s, or a stiff or dense soil condition where the soil depth is less than 200 ft
(b) $S = 1·2$, for dense or stiff soil conditions where the soil depth exceeds 200 ft
(c) $S = 1·5$, for soft to firm clays or loose sands 30 ft or more in depth.

### 6.4.5. Structural type factor K

The inherent ductile nature, redundancy and damping of a building structure are of great importance to its good performance in an earthquake. Factors contributing to this are material and member ductility, a high degree of redundancy with respect to all failure modes, regular form, low eccentricity, good construction quality control and high damping.

Customarily $K$ factors for buildings as a whole vary so that the highest value is twice the lowest, but parts of buildings such as parapets, towers, tanks, chimneys and other appendages may be assigned much higher values.

The Building Seismic Safety Council (1985) and the Structural Engineers Association of California (1985) both use an alternative approach to $K$ factors, having instead $R$ factors which are divisors instead of multipliers in the lateral force calculation.

### 6.4.6. Weight W

The weight $W$ is normally the total dead load plus an estimate of the possible live load that could reasonably be expected.

### 6.4.7. Period T

Because the design loading depends on the building period, and the period cannot be calculated until a design has been prepared, most codes provide formulae from which $T$ may be calculated. The International Conference of Building Officials (1985) gives the building period $T$ in seconds for moment frames as

$$T = 0·1N \tag{6.1}$$

and for stiff buildings (shear wall, braced frame) as

$$T = 0·5h_n/D^{1/2} \tag{6.2}$$

(feet units) or

$$T = 0·9h_n/D^{1/2} \tag{6.3}$$

95

(metre units), where $N$ is the number of storeys, $h_n$ is the building height and $D$ is the building dimension parallel to the seismic force.

The Structural Engineers Association of California (1985) gives alternative formulae for building period $T$ for stiff buildings. Equations (6.2) and (6.3) are used but $D$ is redefined as

$$D = D_{smax} \sum \frac{D_s^2}{D_{smax}^2} \tag{6.4}$$

where $D_s$ is the length of shear wall or braced frame segments in a direction parallel to the applied forces, $D_{smax}$ is the length of the longest shear wall or braced frame segment and $D$ need not exceed the dimensions of the structure in the direction parallel to the applied forces.

For moment frames the Structural Engineers Association of California (1985) gives

$$T = 2 \cdot 44 C_t h_n^{0 \cdot 75} \tag{6.5}$$

(feet) or

$$T = C_t h_n^{0 \cdot 75} \tag{6.6}$$

(metres), where $C_t = 0 \cdot 035$ for steel frames and $C_t = 0 \cdot 030$ for concrete frames.

### 6.4.8. Distribution of lateral forces

Using the parameters described a base shear can be calculated in accordance with the appropriate design code. The distribution of forces making up this shear depends on the dynamic response shape of the building. The most common formula for lateral force distribution is

$$f_i = \frac{V w_i h_i}{\sum_i w_i h_i} \tag{6.7}$$

where $V$ is the base shear, $w_i$ is the mass at level $i$, $h_i$ is the height above ground of level $i$ and $f_i$ is the force at level $i$.

The Applied Technology Council (1982) introduced a more accurate approach which reflected the influence of higher models as

$$f_i = \frac{V w_i h_i^k}{\sum_i w_i h_i^k} \tag{6.8}$$

where the value of $k$ is related to the building period $T$ as follows

(a) $k = 1$ for buildings where $T$ is less than $0 \cdot 5$ s
(b) $k = 2$ for buildings having a period of $2 \cdot 5$ s or greater
(c) $k = 1\text{–}2$, linearly interpolated for $T = 0 \cdot 5\text{–}2 \cdot 5$ s.

Newmark & Hall (1982) give a method for checking whether the lateral force distribution is in agreement with the structure as designed. The procedure is as follows.

(a) Calculate lateral forces $f_i$ according to equation (6.8).

(b) Select member sizes for the structure.

(c) Compute lateral displacements $x_i$ for the structure under the action of the lateral forces from step (a).

(d) Calculate new lateral forces from equation (6.8), replacing the values of $h_i^k$ with the computed values of $x_i$.

(e) Compare the recomputed storey shears with the values derived from step (a). If any of these differ by more than 30% a dynamic analysis should be undertaken.

This method is a useful check on the assumption of regularity on which the approximate formulae for force distribution are based. Generally the alternative method of dynamic analysis will be by response spectrum modal analysis.

An example of lateral force calculation is given in Section 6.8.

## 6.5. Designing for ductility
### 6.5.1. Principles

Ductility is achieved in structural members firstly by designing elements within known limits where they can deform in a ductile manner and secondly by avoiding any possibility of brittle failure. Some of the possible types of brittle failure are given in Table 6.1.

Avoiding the possibility of brittle failure means that at *ultimate load conditions* there is still an adequate safety margin between the actual stress and the brittle failure stress. For example a tension bolt in a steel beam connection should be at a safe stress level when the beam has reached its ultimate moment.

Designing whole structural systems for ductility requires

(a) any mode of failure should involve the maximum possible redundancy

(b) brittle-type failure modes, such as overturning, should be adequately safeguarded so that ductile failure will occur first.

*Table 6.1. Types of brittle failure*

| | |
|---|---|
| Structure | Overturning |
| Foundation | Rotational shear failure |
| Structural steel | Bolt shear or tension failure |
| | Member buckling |
| | Member tension failure |
| | Member shear failure |
| | Connection tearing |
| Reinforced concrete | Bond or anchorage failure |
| | Member tension failure |
| | Member shear failure |
| Masonry | Out-of-plane bending failure |
| | Toppling |

### 6.5.2. Ductility

The definition of ductility factor is shown in Fig. 6.2 and an approach to estimating the ductility requirement of a structure is given in Chapter 3. It is generally accepted that by following a modern design code sufficient ductility will be provided.

However, and this cannot be over-emphasised, design codes are prepared for regular structures. The type of structure where high ductility may be needed is one where a light flexible structure is attached to a larger structure, e.g. a roof tank or penthouse, and such a configuration may require careful analysis.

## 6.6. Other code structural requirements

### 6.6.1. Drift

Drift, or interstorey displacement, should be limited as it is the prime source of non-structural damage and human discomfort and leads to secondary stresses in the main structure. It is normally specified at the elastic design level although it will be greater for the maximum earthquake. Code values vary between 0·002 and 0·008 times the storey heights, with 0·005 being in common use.

### 6.6.2. Vertical acceleration

Specific values are not generally given in codes, although vertical coefficients of 1·2 and 1·4 are used in Italy and France. In general buildings are not particularly susceptible to vertical ground motion but its effect should be borne in mind in the design of reinforced concrete columns, steel column connections and prestressed beams. A specific consideration of vertical ground motion should be given to long span structures and cantilevers should be designed for possible reverse bending.

A common approximate value given to vertical acceleration is a ratio of 0·67 to the horizontal acceleration, but this should be treated as a possible median value and certainly not used for sensitive structures.

### 6.6.3. Horizontal torsion

Torsion arises from the building form and from torsional ground motion. American, Canadian and Mexican codes deal with the ground motion component by increasing the eccentricity of the lateral design force. A common assumption in codes is that the lateral force acts with an eccentricity from the centre of mass equal to 5% of the building base dimension, measured at right angles to the direction of the applied force.

Large eccentricities should be investigated by dynamic analysis.

### 6.6.4. Overturning

Overturning moments calculated by the design lateral forces may be over-conservative, owing to the dynamic simplification made in their derivation. The Canadian code provides reduction factors for stresses arising from overturning, but the 'J' factor, serving a similar function, introduced into early versions of the Structural Engineers Association of California code was withdrawn in 1970.

Some codes require that only 75% of the dead load be mobilised in resisting overturning to take account of vertical acceleration.

### 6.6.5. *Combination of directions*

Codes vary on the need to combine horizontal seismic forces in orthogonal directions. There is no reason to expect earthquake forces to apply themselves along building axes, so that the maximum acceleration could be assumed to occur in any direction. However, there is no reason why the orthogonal components should be in phase. A reasonable compromise may be to follow an approach used in some codes of combining 100% of the code lateral force on one axis with 30% of the appropriate force in a direction at 90°.

## 6.7. Combination of forces

The manner in which the forces arising from different categories of loading are combined varies from code to code and depends on the way in which the loads and stress levels are defined. For this reason it is always preferable to follow a consistent approach. However, some typical code rules for load combination are quoted here for general guidance

$$U = 1{\cdot}0D + 1{\cdot}3L + 1{\cdot}0E \tag{6.9}$$

$$U = 0{\cdot}9D + 1{\cdot}0E \tag{6.10}$$

$$U = 1{\cdot}0D + S + 1{\cdot}0E \tag{6.11}$$

(NZS 3101:1982)

$$U = 1{\cdot}05D + 1{\cdot}6L + 1{\cdot}4E \tag{6.12}$$

(American Concrete Institute, 1983)

$$U = (1{\cdot}1 + 0{\cdot}5A_v)D + 1{\cdot}0L + 1{\cdot}0S + 1{\cdot}0E \tag{6.13}$$

$$U = (0{\cdot}9 - 0{\cdot}5A_v)D + 1{\cdot}0E \tag{6.14}$$

for brittle elements

$$U = (0{\cdot}7 - 0{\cdot}5A_v)D + 1{\cdot}0E \tag{6.15}$$

(Building Seismic Safety Council, 1985) ($A_v$ is the value of the effective peak velocity-related acceleration, taken as zero for values below $0{\cdot}05g$). Here $U$ is the ultimate strength, $D$ is the effect of a dead load, $L$ is the effect of a live load, $S$ is the effect of a snow load and $E$ is the effect of an earthquake load.

Live load is the load applicable in combination with the earthquake load, and the combination with snow load in NZS 3101:1982 is for areas where snow loading can be expected to be sustained over a considerable period. In all cases the effect of the earthquake load should be taken from any direction and the orthogonal combination discussed in Section 6.6.5 allowed for.

## 6.8. Example of lateral force calculation

Using the building structure defined in Chapter 3, Table 3.6, the following calculation of the base shear, using the *Uniform building code* (International Conference of Building Officials, 1985), applies

$$V = ZICKSW$$

$$= 1 \times 1 \times 0.67 \times 0.8 \times 1 \times 1.225 \times 10^6 \times 9.81 \text{ N}$$

$$= 6.44 \text{ MN}$$

Using equation (6.8), $k = 1.25$, the storey forces $f_i$ and the corresponding interstorey shears $v_i$ are derived as given in Table 6.2.

Using the displacements calculated for the building in Chapter 3 (Section 3.15.2) the distribution of forces can be reviewed as given in Table 6.3. The largest discrepancy between $v_i'$, derived from calculated displacements, and $v_i$, calculated from equation (6.8), is 11%.

Table 6.2. *Storey forces and interstorey shears*

| Level | Mass: $\times 10^6$ kg | Height: m | $w_i h_i$: $\times 10^{-6}$ | $f_i$: kN | $v_i$: kN |
|---|---|---|---|---|---|
| 1 | 1·225 | 3·0 | 4·8 | 72 | 6440 |
| 2 | 1·225 | 6·0 | 11·5 | 173 | 6368 |
| 3 | 1·225 | 9·0 | 19·1 | 288 | 6195 |
| 4 | 1·225 | 12·0 | 27·4 | 413 | 5907 |
| 5 | 1·225 | 15·0 | 36·2 | 546 | 5494 |
| 6 | 1·225 | 18·0 | 45·4 | 684 | 4948 |
| 7 | 1·225 | 21·0 | 56·3 | 849 | 4264 |
| 8 | 1·225 | 24·0 | 65·1 | 981 | 3415 |
| 9 | 1·225 | 27·0 | 75·4 | 1137 | 2434 |
| 10 | 1·225 | 30·0 | 86·0 | 1297 | 1297 |
| Column total | | | 427·2 | 6440 | |

Table 6.3. *Distribution of forces*

| Level | $x_i$: $\times 10^{-6}$ mm | $w_i x_i$ | $f_i'$: kN | $v_i'$ kN | $v_i'/v_i$: |
|---|---|---|---|---|---|
| 1 | 1·1 | 1·35 | 66 | 6440 | 1·0 |
| 2 | 2·3 | 2·82 | 139 | 6374 | 1·0 |
| 3 | 3·4 | 4·17 | 205 | 6235 | 1·01 |
| 4 | 6·0 | 7·35 | 362 | 6030 | 1·02 |
| 5 | 8·4 | 10·3 | 507 | 5668 | 1·03 |
| 6 | 10·6 | 13·0 | 640 | 5161 | 1·04 |
| 7 | 12·6 | 15·4 | 758 | 4521 | 1·06 |
| 8 | 17·7 | 21·7 | 1068 | 3763 | 1·1 |
| 9 | 21·4 | 26·2 | 1290 | 2695 | 1·11 |
| 10 | 23·3 | 28·5 | 1405 | 1405 | 1·08 |
| Column total | | 130·79 | 6440 | | |

## 6.9. Bibliography

Berg, G. V. (1982). *Seismic design codes and procedures.* Berkeley: Earthquake Engineering Research Institute.

Building Seismic Safety Council (1985). *NEHRP recommended provisions for the development of seismic regulations for new buildings.* Washington DC: Building Seismic Safety Council.

Earthquake Engineering Research Institute (1987). Seismic building codes: status and prospects. *Annual Meet. San Diego.*

International Association for Earthquake Engineering (1984). *Earthquake resistant regulations: a world list.* Distributed by Gakujutsu Bunken Fukyu-Kai, Oh-Okayama, 12-1, Meguroku, Tokyo, 152, Japan.

International Conference of Building Officials (1985). *Uniform building code,* 1985 edn. Obtainable from 5360 South Workman Mill Road, Whittier, CA 90601, USA.

Newmark, N. M. & Hall, W. J. (1982). *Earthquake spectra and design.* Berkeley: Earthquake Engineering Research Institute.

Structural Engineers Association of California (1974). *Recommended lateral force requirements and commentary.* Obtainable from SEAOC, PO Box 19940, Sacramento, CA 95819-0440, USA.

Structural Engineers Association of California (1985). *Tentative lateral force requirements, October 1985.* Obtainable from SEAOC, PO Box 19940, Sacramento, CA 95819-0440, USA.

## Chapter 7

# Reinforced concrete design

'It is necessary to have an understanding of
the manner in which a structure absorbs the
energy transmitted to it by an earthquake.'
*Design of multistorey reinforced concrete
buildings for earthquake motions*, J. A. Blume,
N. M. Newmark and L. H. Corning, 1961

The scope this chapter covers

(*a*) the lessons from earthquake damage
(*b*) ductility in reinforced concrete members
(*c*) design and detailing of beams
(*d*) design and detailing of columns
(*e*) beam–column joints
(*f*) design and detailing of shear walls
(*g*) seismic provisions in slabs
(*h*) seismic provisions in prestressed concrete
(*i*) seismic provisions in precast concrete.

## 7.1. Introduction

The use of reinforced concrete as a ductile material began in the
early 1960s with the publication of Blume, Newmark & Corning (1961)
which established that properly detailed reinforced concrete beams
and columns would respond to dynamic forces in a ductile manner
and would sustain a number of cycles of stress reversal. The same
conclusion was later drawn for shear walls, principally the work of
Professor R. Park and Professor T. Paulay at the University of Can-
terbury in New Zealand during the 1970s.

Design codes for reinforced concrete in seismic zones are well
established and when properly applied provide a sound basis for
design and detailing. It is unusual in design practice for a calculation
of either the ductility capacity or the ductility requirement to be
necessary. Only very irregular structures are likely to have a deficit
of ductility if good detailing practice is followed. The reader's attention is drawn to comments in the introduction to this book on the use
of recommendations from more than one source, which are particularly relevant to this chapter.

Some codes make provisions for designing for limited ductility and
this is reflected in the reinforced concrete detailing requirements.
The basis of this method is to use higher design forces and to provide

lower levels of ductility. Because the resulting detailing of reinforced concrete, which is what this chapter is concerned with, then becomes considerably less complex, this method is not dealt with here. Codes providing for this approach include NZS 3101:1982 and the *CEB model code for seismic design of concrete structures*, 1985.

A similar argument applies to design in areas of moderate or low seismicity. Most codes provide for lower ductility and consequently less stringent detailing requirements where lower intensities of ground shaking can be expected.

## 7.2. Lessons from earthquake damage

The principal forms of damage in reinforced concrete elements are described in Chapter 1. Once there is a loss of integrity in structural elements, mechanisms of overall or partial collapse can occur.

The need to learn from earthquake damage studies and to apply good engineering sense and judgement based on this learning cannnot be emphasised too highly. It is far more important than any amount of computation and analysis. The common sense lessons from damage studies are as follows.

(a) All frame elements must be detailed so that they can respond to strong earthquakes in a ductile fashion. Any elements which are necessarily incapable of ductile behaviour must be designed to remain elastic at ultimate load conditions.

(b) Non-ductile modes such as shear and bond failures must be avoided. This implies that the anchorage and splicing of bars should not be done in areas of high concrete stress, and a high resistance to shear should be provided.

(c) Rigid elements should be attached to the structure with ductile or flexible fixings.

(d) A high degree of structural redundancy should be provided so that as many zones of energy-absorbing ductility as possible are developed before a failure mechanism is created. For framed structures this means that the yielding should occur first in the beams. Failure in columns should be avoided; they should remain elastic at the maximum design earthquake level.

(e) Joints should be provided at discontinuities, with adequate provision for movement so that pounding of the two faces against each other is avoided.

## 7.3. Materials

The minimum concrete strength generally considered suitable for seismically resistant structures is 20 N/mm² (3000 lbf/in²). ACI 318–83 (American Concrete Institute, 1983) imposes an upper limit for lightweight concrete of 27 N/mm² (400 lbf/in²), unless a higher value can be justified by testing, on the grounds that it tends to be a more brittle material than normal concrete.

For reinforcement the provision, firstly, of adequate ductility and, secondly, of an upper limit on the yield stress, or characteristic strength, are essential. General practice is to limit the yield stress of reinforcement to 410 N/mm$^2$ (60 000 lbf/in$^2$).

## 7.4. Ductility in reinforced concrete members

Ductility, also discussed in Chapter 3, can be defined as the ratio of displacement at maximum load to displacement at yield. Two values of ductility are of concern. Firstly the ductile capacity is the value at ultimate member load, and secondly the ductility requirement is the value at the ultimate design load.

For practical values of section size and reinforcement, section ductile capacity is increased for

(a) an increase in compression steel content
(b) an increase in concrete compressive strength
(c) an increase in ultimate concrete strain.

Similarly section ductile capacity is decreased for

(a) an increase in tension steel content
(b) an increase in steel yield strength
(c) an increase in axial load.

The effect of confining concrete with stirrups or spiral reinforcement is to increase the ultimate concrete strain, thereby increasing the ductile capacity. There is a further advantage in practice as shear resistance is increased and additional lateral support is given to the main reinforcement. Practical values of stirrups or spiral reinforcement which will provide effective containment are substantially larger than those customarily used for reinforced concrete design in non-seismic conditions. Fig. 7.1 illustrates the effects of axial load and confinement on rotational ductile capacity.

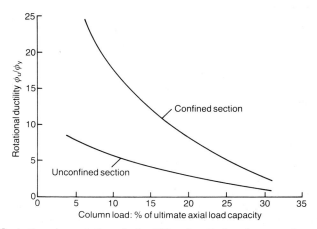

*Fig. 7.1. Variation in rotational ductility for tied columns of confined and unconfined concrete in relation to axial load* ($p = A_s/bh = 0.05$; $f_c' = 3000$ *lbf/in$^2$*; $f_y = 40\,000$ *lbf/in$^2$*; $f_y'' = 60\,000$ *lbf/in$^2$*)

A detailed discussion of the quantitative effects of the various parameters on ductile capacity is given in Park & Paulay (1975).

## 7.5. Reinforced concrete beams

### 7.5.1. Dimensions

Limiting dimensions for beams are given in ACI 318–83 (American Concrete Institute, 1983) as

(a) $b/d$ not less than 0·3
(b) $b$ not less than 250 mm
(c) $b$ not greater than the column width plus 0·75$d$ on each side.

In NZS 3101:1982 the following rules are given

(a) $b$ not less than $L/25$, $Ld/100$, 200 mm
(b) for cantilevers, $b$ not less than $L/15$, $Ld/60$, 200 mm
(c) for T and L beams, $b$ not greater than $L/16·7$
(d) for cantilever T and L beams, $b$ not greater than $L/10$.

Here $b$ is the beam width, $L$ is the beam span and $d$ is the beam depth.

### 7.5.2. Reinforcement

*7.5.2.1. Main bar reinforcement.* For main bars an upper limit for the reinforcement ratio of 0·025 is established, above which experiment has shown that there is inadequate ductility. Also, on practical grounds, values above 0·025 result in considerable congestion. A lower limit is also provided of $1·4/f_y$ (N/mm$^2$) or $200/f_y$ (lbf/in$^2$). The minimum bar diameter permissible is 12 mm and there must be at least two bars in both the top and the bottom face. The minimum diameter for stirrups is 10 mm.

*7.5.2.2. Hinge zone.* Within the zone of a potential plastic hinge additional requirements exist. Normally the hinge occurs at the end of the beam and this zone is defined as extending from the end of the beam for a distance of twice the beam depth. In order to provide adequate ductility in the hinge zone, the following additional requirements are given in NZS 3101:1982.

(a) The tension reinforcement ratio should not exceed the values given by

$$\rho_{max} = \frac{1 + 0·17(f_c'/7 - 3)}{100}\left(1 + \frac{\rho'}{\rho}\right) \qquad (7.1a)$$

$$\rho_{max} = 7/f_y \qquad (7.2a)$$

with $f_y$ in newtons per millimetre squared. For $f_y$ in pounds force per square inch the equations are

$$\rho_{max} = \frac{1 + 0·17(f_c'/1000 - 3)}{100}\left(1 + \frac{\rho'}{\rho}\right) \qquad (7.1b)$$

$$\rho_{max} = 1000/f_y \qquad (7.2b)$$

where $\rho$ is the ratio of tension reinforcement, $\rho'$ is the ratio of compression reinforcement, $f_y$ is the reinforcement yield stress and $f_c'$ is the specified concrete cylinder compressive strength.

(b) A minimum of two 16 mm bars should be provided in both top and bottom faces.

(c) The stirrup spacing should not exceed $d/4$, six times the diameter of the largest main bar restrained or 150 mm. The first stirrup tie must be within 50 mm of the column face.

(d) Normally stirrups complying with the foregoing and designed to provide adequate shear resistance will provide sufficient anchorage to the main bars. They should generally be arranged to give the most effective anchorage possible against buckling of the main reinforcement in the confined zone. Additional requirements are laid down in NZS 3101:1982 for restraint against buckling.

*7.5.2.3. Bending moments.* In order to allow for uncertainty in the performance of the structure in an earthquake, minimum bending moments are specified as follows.

(a) The positive moment strength at the joint face shall be not less than one-half of the negative moment strength provided.

(b) Neither the negative nor the positive moment strength at any point in the member shall be less than a quarter of the maximum moment provided at the face of either joint.

*7.5.2.4. Splices.* Lap splices of main bars should be located as far as possible in zones of low stress. Splices are not acceptable within the column zone nor within zones of a potential plastic hinge. Where splices occur stirrups should be spaced at a maximum of a quarter of the beam depth or 100 mm. Lap splices are not permitted within joints nor in areas of possible flexural yielding.

*7.5.2.5. Beams.* In calculating the ultimate moments of beam sections in the case of T or L beams, account will have to be taken of any slab reinforcement which may participate. NZS 3101:1982 lays down rules for this as follows

(a) at interior columns with a transverse beam, all reinforcement within the slab up to four times the slab thickness on each side of the beam

(b) at interior columns without a transverse beam, all reinforcement within the slab up to 2·5 times the slab thickness on each side of the beam

(c) at exterior columns with a transverse beam, all reinforcement within twice the slab thickness on each side of the column

(d) at exterior columns with no transverse beam, all reinforcement within the width of the column.

Typical arrangements of beam reinforcement are shown in Figs 7.2 and 7.3.

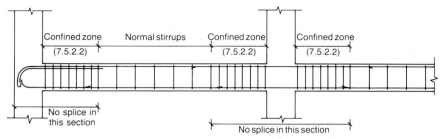

Fig. 7.2. Typical beam reinforcement (numbers in parentheses refer to the sub-sections in the text): minimum areas of steel based on moment capacity and the minimum diameter of the main bars are given in Sections 7.5.2.1 and 7.5.2.2; the minimum diameter of the stirrups is given in Section 7.5.2.1; minimum bending moments are given in Sections 7.5.2.3 and 7.5.2.5; splices are described in Section 7.5.2.4

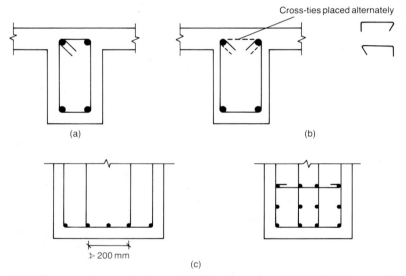

Fig. 7.3. Beam section and stirrup details: (a) preferred closed stirrups with 10 dia. 135° bends; (b) alternative stirrup with cross-ties; (c) stirrups arranged to provide anchorage to the main bars (beam dimensions are given in Section 7.5.1)

## 7.6. Reinforced concrete columns

### 7.6.1. Dimensions

ACI 318–83 (American Concrete Institute, 1983) gives the minimum column dimension as 300 mm (12 in) whereas NZS 3101:1982 gives 200 mm (8 in). Although stability and strength requirements will usually require a larger section than 200 mm it is difficult to see how a section depth of less than 300 mm can be effective in bending. Additionally ACI 318–83 gives a limiting ratio to the column face dimensions of 0·4.

### 7.6.2. Reinforcement

*7.6.2.1. Reinforcement ratio.* ACI 318-83 requires that the reinforcement ratio for longitudinal bars lies between 0·01 and 0·06. NZS 3102:1983 gives a minimum value of 0·008 with maximum values of 0·06 for grade 275 steel ($f_y = 275$ N/mm$^2$) and 0·045 for grade 380 steel ($f_y = 380$ N/mm$^2$). At splices the maximum ratio is increased by one-third.

*7.6.2.2. Minimum bar diameters.* Minimum bar diameters are generally accepted as

(a) longitudinal bars, 12 mm
(b) stirrups, 10 mm
(c) supplementary cross-ties, 8 mm.

*7.6.2.3. Longitudinal bars.* Longitudinal bars in the plastic hinge region should not be spaced more than 200 mm apart, and the smallest bar diameter used should not be less than two-thirds of the largest bar diameter in any one row.

*7.6.2.4. Column bars.* NZS 3102:1983 provides maximum sizes for column bars traversing intersecting beams, in order to avoid bond failure within the joint. Where plastic hinges may occur the maximum diameter of such bars is 1/20th of the beam depth for grade 275 steel ($f_y = 275$ N/mm$^2$) and 1/25th of the beam depth for grade 380 steel ($f_y = 380$ N/mm$^2$). Where it can be shown that plastic hinges will not occur these values are reduced to 1/15th and 1/20th of the beam depth respectively.

*7.6.2.5. Splices.* Lap splices are limited to the middle half of the column by ACI 318–83 and by NZS 3101:1982 to the middle quarter unless it can be shown that plastic hinges cannot occur. All splices in columns should be full tension splices. There are considerable practical difficulties in implementing the mid-height splicing of column reinforcement, although as long as plastic hinges can develop at the ends of a column the need for this requirement is obvious. Alternative solutions using mechanical or welded splices can be adopted and regulations covering these are given in NZS 3101:1982.

*7.6.2.6. Transverse reinforcement.* Rules for the provision of transverse reinforcement in the plastic hinge zone of columns are given in NZS 3102:1982 and ACI 318-83. These rules differ substantially and both sets are given here.

ACI 318-83 requires a stirrup spacing not greater than

(a) a quarter of the least column dimension
(b) 100 mm.

The spacing of stirrup legs or cross-ties on plan should not exceed 350 mm. The volumetric ratio of stirrups is given by

$$\rho_s = 0\cdot12f_c'/f_{yh} \tag{7.3}$$

where $f_c'$ is the concrete compressive strength and $f_{yh}$ is the yield strength of the stirrups.

109

The total cross-sectional area of stirrups shall not be less than either of

$$A_{sh} = 0.3 \frac{sh_c f_c'}{f_{yh}} \left( \frac{A_g}{A_{ch}}^{-1} \right)$$ (7.4)

$$A_{sh} = 0.12 \frac{sh_c f_c'}{f_{yh}}$$ (7.5)

where $A_{sh}$ is the total cross-sectional area of transverse reinforcement (including cross-ties) within spacing $s$ and perpendicular to dimension $h_c$, $A_g$ is the gross cross-sectional area of column and $A_{ch}$ is the cross-sectional area measured from outside to outside of the transverse reinforcement.

*7.6.2.7. Distance for transverse reinforcement.* The distance over which the transverse reinforcement is required is the greater of

(a)  the column depth
(b)  one-sixth of the clear column height
(c)  450 mm.

*7.6.2.8. Stirrup spacing.* NZS 3102:1982 requires a stirrup spacing not greater than

(a)  one-fifth of the least column dimension
(b)  six times the diameter of the largest bar restrained
(c)  200 mm.

*7.6.2.9. Volumetric ratio and cross-sectional area.* NZS 3101:1982 requires that the volumetric ratio of stirrups shall not be less than

$$\rho_s = 0.45 \left( \frac{A_g}{A_c} - 1 \right) \frac{f_c'}{f_{yh}} \left( 0.5 + 1.25 \frac{P_e}{\phi f_c' A_g} \right)$$ (7.6)

$$\rho_s = 0.12 \frac{f_c'}{f_{yh}} \left( 0.5 + 1.25 \frac{P_e}{\phi f_c' A_g} \right)$$ (7.7)

The cross-sectional area of stirrup shall be not less than

$$A_{sh} = 0.3 s_h h'' \left( \frac{A_g}{A_c} - 1 \right) \frac{f_c'}{f_{yh}} \left( 0.5 + 1.25 \frac{P_e}{\phi f_c' A_g} \right)$$ (7.8)

$$A_{sh} = 0.12 s_h h'' \frac{f_c'}{f_{yh}} \left( 0.5 + 1.25 \frac{P_e}{\phi f_c' A_g} \right)$$ (7.9)

where $A_{sh}$ is the area of stirrups and cross-ties, $s_h$ is the stirrup spacing, $h''$ is the dimension of the concrete core (to the outside of the stirrups), $A_g$ is the gross area of the concrete section, $A_c$ is the area of the concrete core (to the outside of the stirrups), $f_c'$ is the concrete cylinder's compressive strength, $f_{yh}$ is the yield strength of the stirrup, $P_e$ is the design axial load and $\phi$ is the strength reduction factor (0.9).

*7.6.2.10. Length.* The length over which the transverse reinforcement will be required varies between 1·0 and 1·5 times the larger column dimension depending on the loading condition.

*7.6.2.11. Restraint.* Generally it will be necessary to provide multiple stirrups, or stirrups and cross-ties, in order to give satisfactory confinement and restraint to main column reinforcement. Generally overlapping hoops are to be preferred, cross-ties being less efficient in restraint against buckling. In either case one stirrup should surround the whole of the main reinforcement. Where restrained main bars are less than 200 mm apart, it is not necessary to restrain intermediate bars.

The provision of lateral restraint is based on the assumption that the restraining bar should provide a force equal to 1/16th of the force in the main bar or bars restrained.

Figures 7.4 and 7.5 show typical reinforcement details for columns.

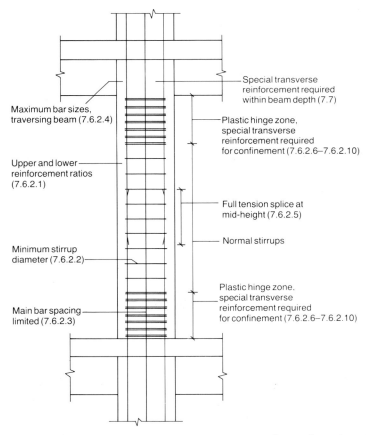

*Fig. 7.4. Column detailing (numbers in parentheses refer to the subsections in the text)*

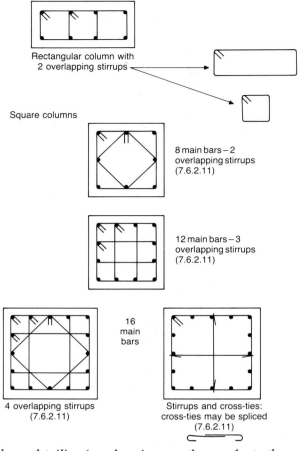

*Fig. 7.5. Column detailing (numbers in parentheses refer to the subsections in the text) (the limiting aspect ratio and dimension is given in Section 7.6.1)*

## 7.7.  Beam–column joints
### 7.7.1.  *Internal beam–column joint*
Figure 7.6 shows the forces acting on an internal beam–column joint. Damage studies have shown that considerable distress may be suffered in this area, the principal mechanisms of failure being

(*a*)  shear within the joint
(*b*)  anchorage failure of beam bars anchored in the joint
(*c*)  bond failure of beam or column bars passing through the joint.

### 7.7.2.  *Effects of loading*
It should be emphasised that these effects exist under non-seismic loading but are of much greater significance under seismic conditions, and the effects are aggravated by cyclic loading. Design prac-

112

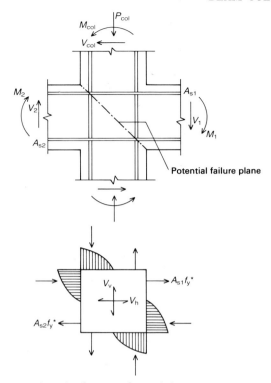

*Fig. 7.6. Forces on an interior beam–column joint*

tice for joints, which reflects current knowledge, is given in detail in NZS 3101:1982. Practice is based on the fundamental concept that failure should not occur within the joint, so that it should be strong enough to withstand the yielding of connecting beams (usually) or columns.

### 7.7.3. Shear resistance

Horizontal shear strength within the joint can be calculated for non-prestressed reinforced concrete by

$$V_h = (A_{s1} + A_{s2})f_y^* - V_{col} \tag{7.10}$$

where the symbols are those shown in Fig. 7.6 and $f_y^*$ is a factored yield strength which allows for overstrength. For $f_y = 275$ N/mm$^2$, $f_y^* = 344$ N/mm$^2$, and for $f_y = 380$ N/mm$^2$, $f_y^* = 532$ N/mm$^2$.

The shear $V_h$ is resisted by compressive strut action in the concrete and horizontal stirrups, as shown in Fig. 7.7. Guidance on the value of strut action should be sought from the code referred to, but, conservatively, the required area of horizontal stirrups can be calculated from

$$A_{sh} = V_h/f_y \tag{7.11}$$

113

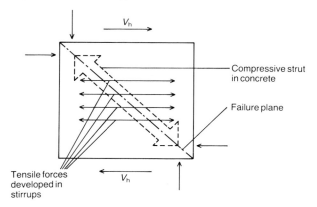

*Fig. 7.7. Horizontal shear resistance in a beam–column joint*

A minimum cross-sectional area of horizontal stirrups is recommended by the American Concrete Institute–American Society of Civil Engineers Committee 352 (1985) as not less than either of

$$A_{sh} = 0 \cdot 3 \, \frac{s_{h} h'' f_{c}'}{f_{yh}} \left( \frac{A_{g}}{A_{c}} - 1 \right) \tag{7.12}$$

$$A_{sh} = 0 \cdot 09 \, \frac{s_{h} h'' f_{c}'}{f_{yh}} \tag{7.13}$$

where the symbols are those defined following equation (7.9).

Stirrups must cross the failure plane shown in Fig. 7.7 and be anchored at a distance that is not less than one-third of the appropriate column dimension on each side of it. The maximum spacing of stirrups should not exceed that appropriate to the adjacent column. Confinement steel within the joint is the same as for the adjacent column except that, where the joint is confined on all four sides by connecting beams, this reinforcement may be reduced by a half.

### 7.7.4. Anchoring

Unless special measures are taken to remove the plastic hinge region away from the face of the column, the onset of yielding in the beam will penetrate the column area. For this reason the anchorage length of beam bars anchored within the column area on external joints is reduced by the lesser of half the column depth or 10 times the bar diameter as illustrated in Fig. 7.8. One solution to the difficult problem of anchoring beam bars in external joints is the use of beam stubs as shown in Fig. 7.9.

### 7.7.5. Bond stresses

At interior beam–column joints high bond stresses may develop, where bars may have a high tensile stress at one face and a high

114

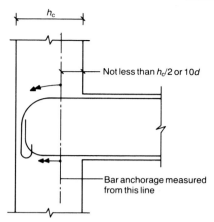

*Fig. 7.8. Anchorage of beam bars in an external joint*

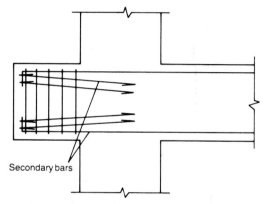

*Fig. 7.9. Use of a beam stub for anchoring main bars in an external joint*

compressive stress on the other face, and yield penetration from the plastic hinge in the beam may reduce the effective bond. The New Zealand code limits bar diameters in this situation to 1/25th of the column depth for grade 275 steel ($f_y = 275$ N/mm$^2$) or 1/35th for grade 380 steel ($f_y = 380$ N/mm$^2$).

### 7.7.6. Shear stresses

ACI 318-83 (American Concrete Institute, 1983) provides limiting shear stresses for joints in order to avoid the possibility of failure in the compression strut portion, as follows

   (a)  for a confined joint: $240A_j f_c'^{1/2}$ (N/mm$^2$) or $20A_j f_c'^{1/2}$ (lbf/in$^2$)
   (b)  for others: $180A_j f_c'^{1/2}$ (N/mm$^2$) or $15A_j f_c'^{1/2}$ (lbf/in$^2$)

where $A_j$ is the cross-sectional area of the joint and $f_c'$ is the compressive cylinder strength of the concrete.

A joint is regarded as confined if members frame into each face of the joint, each covering at least three-quarters of the joint face area.

Typical joint reinforcement is shown in Figs 7.10–7.13.

## 7.8. Shear walls

### 7.8.1. General

The shear wall (Fig. 7.14), functioning as a large vertical cantilever to resist seismic forces, is an essential element in tall reinforced con-

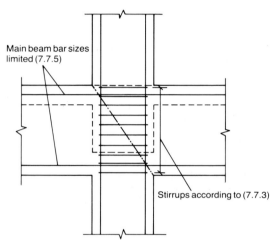

Main beam bar sizes limited (7.7.5)

Stirrups according to (7.7.3)

*Fig. 7.10. Beam–column joint reinforcement (numbers in parentheses refer to the subsections in the text) (for shear stress limitation see Sections 7.7.3 and 7.7.6)*

*Fig. 7.11. Beam–column joint reinforcement in the Financial Complex, Port of Spain: because of high shears both vertical and horizontal stirrups are used in the panel zone (photograph by courtesy of CEP Ltd, Trinidad, published with the permission of the Central Bank of Trinidad and Tobago)*

116

crete structures and a valuable element in those of medium- and low-rise structures. It provides strength more economically than a frame and controls displacements to a degree that cannot be achieved by a frame. It can conform with the architectural requirements for continuous vertical service and lift cores or for plane surfaces on the external faces.

For many years shear walls were regarded as brittle elements. It was assumed that they would behave elastically only for moderate earthquakes. In order to resist major earthquakes they were combined with a ductile frame that was intended to survive after major damage had been inflicted on the shear wall. However, their ability

Fig. 7.12. Beam–column joint and corbel reinforcement in the Financial Complex, Port of Spain (photograph by courtesy of CEP Ltd, Trinidad, published with the permission of the Central Bank of Trinidad and Tobago)

Fig. 7.13. Double-beam stirrup arrangement in a potential hinge zone, in the Financial Complex, Port of Spain (photograph by courtesy of CEP Ltd, Trinidad, published with the permission of the Central Bank of Trinidad and Tobago)

117

*Fig. 7.14. Shear wall reinforcement in a coupling beam in the Financial Complex, Port of Spain (photograph by courtesy of CEP Ltd, Trinidad, published with the permission of the Central Bank of Trinidad and Tobago)*

to respond to major earthquakes in a ductile bending manner is now well established and they can be designed as the sole earthquake-resisting system in a structure, although this is still not permissible in some codes.

The analysis, design and detailing of shear walls is commonly complicated by the necessity to provide openings in regular, or sometimes irregular, patterns.

### 7.8.2. Dimensions

The American Concrete Institute code (1983) ACI 318-83, appendix A, gives minimum reinforcement ratios for reinforced concrete walls, both horizontal and vertical, of 0·0025. The New Zealand code NZS 3101:1982 gives upper and lower limits as follows

(a) mild steel ($f_y = 250$ N/mm² (36 000 lbf/in²)): lower limit, 0·0028; upper limit, 0·064
(b) high yield steel ($f_y = 410$ N/mm² (60 000 lbf/in²)): lower limit, 0·0017; upper limit, 0·039.

These should be interpreted as applying to any portion of the wall. For seismic-resistant construction a layer of reinforcement in each face is required.

The limiting wall dimensions are the same as for normal reinforced concrete construction. Compressive stability considerations at the free edges of walls can be dealt with by treating them as columns.

### 7.8.3. Failure of shear walls

Figure 7.15 shows the main modes in which a shear wall may fail. The possibility of failure by any one of these modes is reduced by

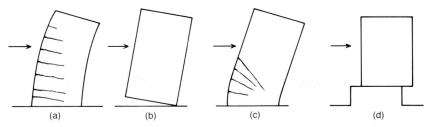

*Fig. 7.15. Principal modes of failure for reinforced concrete shear walls: (a) bending; (b) rocking; (c) diagonal tension; (d) sliding*

increasing the vertical load on the wall, and the common problem in planning is to find a way of increasing the dead load supported by a shear wall. However, the pursuit of dead load should not be taken too far, otherwise the accompanying reduction in ductility may offset the gain in strength.

Where the extreme edges of the wall are not restrained by returns or flanges, the wall may be liable to instability in the plastic hinge area when large strains occur. This is unlikely to occur where the wall thickness at the extreme fibre is greater or equal to 1/10th of the effective length between restraints. This is discussed by Paulay & Goodsir (1985).

Rocking is regarded here as a form of failure although its effects are not as catastrophic as they at first appear and this is discussed in more detail in Chapter 4. Studies of earthquake damage show that sliding failures tend to occur at construction joints so that special treatment at these joints is required.

Once a decision to utilise the ductile capacity of a shear wall has been made, it becomes essential that this controls the response of the structure, and not any other mode of failure. In consequence the wall must be designed to maintain its elastic integrity at all levels other than in the intended ductile zone.

### 7.8.4. Effect of wall shape

Benefit can be obtained in stability and ductility from incorporating flanges in the wall. Fig. 7.16 shows the relative ductilities of flanged and rectangular walls, calculated by Salse & Fintel (1973), for varying values of direct load. Fig. 7.17, from the same source, shows the effect of direct load on ductility and strength for various values of flange reinforcement.

### 7.8.5. Bending strength

As long as the height is greater than twice the depth, the bending strength of a shear wall can be calculated in the same way as that of a beam. This is straightforward where reinforcement is concentrated near to the extreme fibre.

A special case arises where reinforcement is uniformly distributed across the depth of the wall on plan. Cardenas, Hanson, Corbey &

119

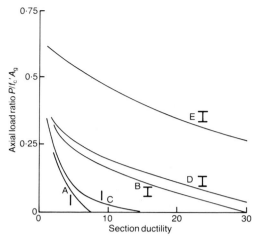

*Fig. 7.16. Effect of cross-section, steel distribution and concrete confinement on ductility (after Salse & Fintel (1970)): curve A, rectangular section, $\rho = 0.02$; curve B, flanged section, $\rho_{\text{flange}} = 0.03$ (curves A and B have the same concentric load capacity); curve C, rectangular section, $\rho = 0.01$; curve D, flanged section, $\rho_{\text{flange}} = 0.01$ (curves C and D have the same pure moment capacity); curve E, flanged section with high confinement steel in the flanges, $\rho_{\text{web}} = 0.01$, $\rho_{\text{flange}} = 0.03$*

Hognestad (1973) have calculated this to be (conservatively)

$$M_u = 0.5 A_s f_y h \left( 1 + \frac{N_u}{A_s f_y} \right) \left( 1 - \frac{c}{h} \right) \qquad (7.14)$$

for any consistent set of units, and where

$$\frac{c}{h} = \frac{\alpha + \beta}{2\beta + 0.85\beta_1}$$

$$\alpha = \frac{1.2 A_s f_y}{b h f_{cu}}$$

$$\beta = \frac{1.2 N_u}{b h f_{cu}}$$

and the symbols are illustrated in Fig. 7.18.

The calculation of shear wall ductility is not simple. A method is given by Salse & Fintel (1973). Generally adherence to design code values and detailing will give more than adequate ductility, and no published design code requires ductility computation for shear walls.

The vertical span of wall over which the hinge zone applies is defined by NZS 3101:1982 as the length on plan or one-sixth of the total height but not more than twice the length on plan. Over this height the main reinforcement needs to be restrained in the same manner as a column.

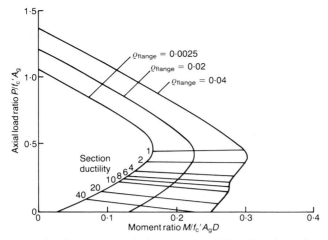

*Fig. 7.17. Direct load, moment, reinforcement and ductility for a flanged wall*

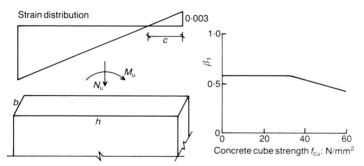

*Fig. 7.18. Ultimate wall strength with uniformly distributed reinforcement (with total area of vertical reinforcement $A_1$ and yield strength of reinforcement $f_y$)*

### 7.8.6. Shear strength

Shear stresses are calculated in the normal way, taking account of the compressive load. In the region of the hinge zone, if the axial compressive stress is less than $0 \cdot 2f_c'$, the contribution of concrete towards the shear strength should be neglected. This takes account of possible adverse combinations of vertical and horizontal accelerations.

In the nominally reinforced upper parts of a shear wall, the provision of minimum reinforcement will give a resistance of $0 \cdot 333f_c'^{1/2}$ N/mm² ($4f_c'^{1/2}$ lbf/in²), which will commonly be sufficient without additional reinforcement. In calculating shear stress the lever arm need not be taken as less than $0 \cdot 8$ times the effective depth.

The shear capacity in combination with direct load may be calculated from ACI 318-83. For compressive loads the shear capacity $V_c$ may be calculated from

$$V_c = \left( 0{\cdot}158f_c'^{1/2} + 17{\cdot}2\rho_w \frac{V_u d}{M_m} \right) b_w d \tag{7.15a}$$

for Newton-millimetre units and from

$$V_c = \left( 1{\cdot}9f_c'^{1/2} + 2500\rho_w \frac{V_u d}{M_m} \right) b_w d \tag{7.15b}$$

for pound-inch units, where $f_c'$ is the concrete compressive strength, $\rho_w$ is the reinforcement ratio based on web thickness, $V_u$ is the shear force at the section, factored, $d$ is the effective depth to the main reinforcement, $b_w$ is the web thickness and $M_m$ is given by

$$M_m = M_u - N_u \frac{(4h - d)}{8} \tag{7.16}$$

where $h$ is the overall depth on plan and $N_u$ is the factored axial load. $V_c$ will be limited to

$$V_c = 3{\cdot}5f_c'^{1/2}b_w d\left( 1 + \frac{N_u}{3{\cdot}45A_g} \right)^{1/2} \tag{7.17a}$$

for Newton-millimetre units and

$$V_c = 42f_c'^{1/2}b_w d\left( 1 + \frac{N_u}{500A_g} \right)^{1/2} \tag{7.17b}$$

for pound-inch units, where $A_g$ is the gross area of the section.
Where there is significant axial tension

$$V_c = 0{\cdot}167\left( 1 + \frac{N_u}{3{\cdot}45A_g} \right)f_c'^{1/2}b_w d \tag{7.18a}$$

for Newton-millimetre units and

$$V_c = 2\left( 1 + \frac{N_u}{500A_g} \right)f_c'^{1/2}b_w d \tag{7.18b}$$

for pound-inch units. $N_u$ is expressed as a negative value for tension.

### 7.8.7. Construction joints
The dependable shear strength of a wall across a construction joint is

$$V_j = \frac{0{\cdot}8N_u + A_v f_y}{0{\cdot}94bh} \tag{7.19}$$

where $N_u$ is the vertical compressive load, $A_v$ is the total area of vertical reinforcement, $f_y$ is the reinforcement's yield stress, $V_j$ is the shear stress, $b$ is the wall thickness and $h$ is the wall depth on plan.

The value of $N_u$ should take account of the possible effect of negative vertical acceleration, reducing it by approximately 20%. The area of reinforcement should not include heavy concentrations near the extremities of the wall as these are not fully effective in providing a clamping action across the joint.

### 7.8.8.  Low-rise shear walls

The behaviour of low-rise shear walls, where the height is less than their length on plan, is not the same as that of taller walls. Park & Paulay (1975) discuss the mechanics of seismic response for these walls and recommend that they are reinforced with bars that are distributed uniformly over their length, with only a nominal increase at the vertical edges.

The steel requirement will usually be satisfied by the minimum content (0·0025) and, although this will provide a relatively low ductility, it will be adequate for the design conditions.

Park & Paulay (1975) also recommend that for the optimum ductile performance of a squat shear wall the shear strength should be limited to $0·5f_c'^{1/2}$ N/mm$^2$ ($6f_c'^{1/2}$ lbf/in$^2$). They point out that the flexural failure mechanism will be accompanied by large cracks, so that the contribution of concrete towards the shear strength should be ignored.

### 7.8.9.  Flanged walls

The effect of flanges is the same as it is on a beam, increasing the compression area and providing additional stability. In cases where there is a substantial compressive load the whole flange and part of the web may be in compression. The design of the compressive portion should then follow practice for column design.

### 7.8.10.  Shear walls with openings and coupled walls

Figure 7.19 shows some typical shear wall arrangements. Each case can be regarded either as a wall with openings or as separate walls coupled with beams. Both damage studies and analysis show that failure will be concentrated around the openings and at the base.

It is necessary to assess both the elastic and the inelastic behaviour of coupled shear walls. Some of the difficulties in elastic analysis are discussed in Chapter 3, and a simplified approach, laminar analysis, for the elastic analysis of coupled walls is described by Park & Paulay (1975).

Figure 7.20 illustrates the critical areas of a coupled shear wall. On the tension side diagonal tension failure may occur, and the vertical tension in the wall reduces the shear capacity at construction joints. On the compression side the direct compressive force is added to bending compression so that high compressive forces occur at the edge. Coupling beams are subjected to a high ductility demand and, if they are made too stiff, may fail in diagonal tension.

123

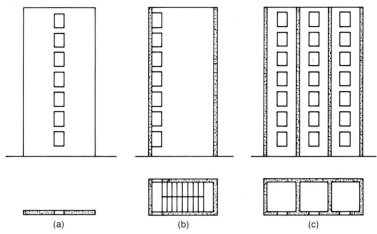

*Fig. 7.19. Typical shear wall arrangements: (a) planar; (b) stair well; (c) lift core*

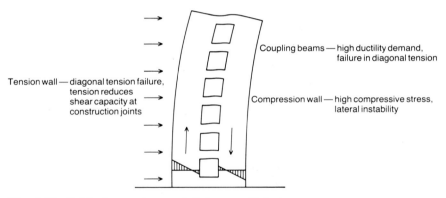

*Fig. 7.20. Critical areas in coupled shear wall design*

The wall design must ensure an appropriate order of yielding and ensure that diagonal tension failure does not occur either in the walls or in the coupling beams. The customary yielding pattern is for plastic hinges to develop first in the coupling beams, followed by hinges at the base of each wall.

Park & Paulay (1975) give criteria for avoiding diagonal tension failure in the coupling beams, by which an upper limit is imposed on the area of main reinforcement

$$\frac{A_s}{bd} \not> \frac{4 \cdot 7 l_s f_c'^{1/2}}{(d - d')f_y} \tag{7.20}$$

where the symbols are illustrated in Fig. 7.21 and $f_c'$ is the concrete

124

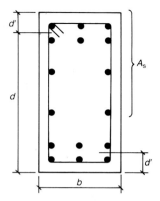

Fig. 7.21. *Coupling beam—section properties*

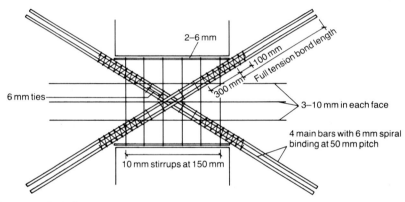

Fig. 7.22. *Reinforcement arrangement in a diagonally reinforced coupling beam (after Park & Paulay (1975))*

cylinder's compressive strength, $f_y$ is the reinforcement's yield strength and $l_s$ is the clear span.

Where the coupling beam depth is greater than half the clear span it becomes difficult to achieve sufficient ductility using the normal beam reinforcement configuration. In this case a reinforcement arrangement, recommended by Park & Paulay (1975) and shown in Fig. 7.22, should be used. The area of main steel in each arm of the X is derived from the ultimate shear $V_u$ and is given by

$$A_s = \frac{V_u}{2f_y \sin \alpha} \tag{7.21}$$

and the ultimate moment of resistance by

$$M_u = V_u l_s/2 \tag{7.22}$$

125

$f_y$ and $l_s$ are the same as for equation (7.20).

Ductilities obtainable from the X configuration have been shown by test to be far superior to those obtainable by conventional reinforcement.

### 7.9. Slabs

The design of slabs in seismic-resistant structures is the same as for non-seismic conditions except in the following particulars.

Slabs function as diaphragms in transmitting forces laterally, especially between vertical elements of differing stiffness. Horizontal shears are thus induced in the slab. For full depth reinforced concrete slabs the shear will generally be insignificant, but for prefabricated or partly prefabricated construction, such as hollow tile slabs, the ability to act as a diaphragm will need to be investigated, particularly where the depth of in situ topping is 50 mm or less.

The use of 'drag bars' may be required where shear walls are only partially embedded in the slab. This is illustrated in Fig. 7.23.

Cantilever slabs should have provision for reverse loading, to take account of a possible combination of negative vertical acceleration with negative bending due to horizontal forces. These slabs function dynamically as appendages in a similar way to roof parapets and are subjected to amplified responses.

Damage studies show that cracking is liable in slabs at discontinuities and a conservative approach should be taken to the detailing of reinforcement at such locations.

### 7.10. Precast concrete

The use of precast concrete in seismic-resistant structures falls into three categories.

The first and most common category is of floor and roof units which do not form part of the moment-resisting structure. The additional requirements over normal use are for secure anchorages at the

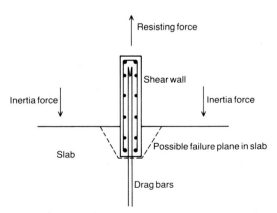

*Fig. 7.23. Provision of drag bars in a slab*

supports and for sufficient lateral connection to be able to function as diaphragms to provide a lateral load path between vertical force-resisting elements. Suitable types of connection between precast slab units are illustrated in Fig. 7.24 and precast slab to beam connections in Fig. 7.25.

The second category is of principal structural elements such as beams, shear walls and slabs. Because the ability to provide ductility at connections is extremely limited it is difficult to design ductile precast systems. In consequence connections will generally be designed on the basis that they are brittle, and ductility provided within the member or within cast-in-place connections. For precast prestressed concrete members it is possible to post-tension connections.

The third category is of cladding elements which have no structural role and are simply appended to the main structure. The design of connections for these units has to satisfy the requirements of

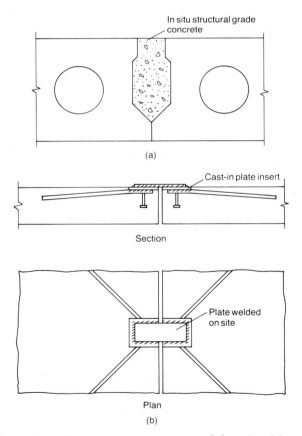

Fig. 7.24. Connections between precast concrete slab units: (a) concrete shear key between precast floor units; (b) welded tie between units

127

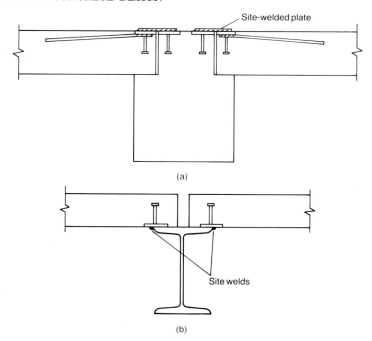

*Fig. 7.25. Precast slab—beam connections: (a) connection of a precast slab to a concrete beam; (b) connection of a precast slab to a steel beam*

strength, convenience in erection, the ability to take up construction tolerances and to provide sufficient flexibility to accommodate movement of the structure without fracture. Suitable types of connection for a precast cladding panel are shown in Fig. 7.26.

For connections generally a comprehensive guide is provided by Martin & Korkosz (1982).

### 7.11. Prestressed concrete
*7.11.1. Introduction*

Prestressed concrete can be designed to provide ductility under cyclic loading and may accordingly be used in the primary seismic-resistant structural system. Nevertheless some caution needs to be exercised when the design conditions make its integrity dependent on the maintenance of gravity loads. In practice the use of prestressed concrete frames for seismic resistance is uncommon and its main use in buildings is for floor and cladding components which are not required to resist seismic forces in a ductile manner.

Prestressed concrete structures subjected to strong ground motion differ from those of normal reinforced concrete in that they have approximately 40% greater displacement, lower damping and lower energy-absorbing capacity in the non-linear range. However, prestressed concrete exhibits a greater elastic recovery so that damage

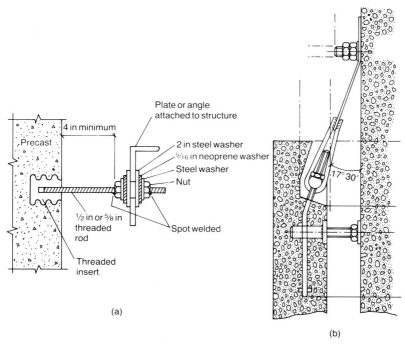

*Fig. 7.26. Precast cladding panel fixings with provision for movement: (a) detail used in Mexico (Fintel & Khan, 1985); (b) Frimeda type 1 cladding panel fixing*

can be expected to be less for moderate earthquakes.

The Fédération Internationale de la Précontrainte (FIP) recommendations (1977, 1978) require design checks at two levels of earthquake

(a) serviceability limit state related to a moderate earthquake
(b) ultimate limit state related to a severe earthquake.

They also refer to considerations of a possible third limit state of the maximum credible earthquake, presumably intended to deal with special structures where the consequences of failure are socially unacceptable, such as nuclear installations. Dealing with the first two categories only, the requirements for the two states are

(a) for the moderate earthquake there must be no loss of prestress: for this check, using elastic theory, the strain in prestressing steel should exceed neither the limit of proportionality nor the strain at transfer
(b) for the severe earthquake, analysis should take into account elasto-plastic deformation and the ultimate limit state and verify that the structure is safe from collapse.

Opinions vary on the use of ungrouted tendons in the seismic force-resisting system. The FIP quotes concern over unbonded tendons on

129

the grounds that failure of the end anchorage will cause a brittle type of failure. Other concerns are the lower energy dissipation characteristics, reduced ductility and difficulty in predicting the ultimate moment capacity under reversed loading. However, NZS 3101:1982 permits the use of unbonded tendons if the prestress is used mainly to balance gravity loads and the seismic resistance is provided by non-prestressed reinforcement. NZS 3101:1982 also recommends the use of corrugated ducts for grouted tendons.

The principles of frame design described in this chapter for reinforced concrete, including capacity design, are equally applicable to prestressed concrete.

### 7.11.2. Damping

Damping of prestressed concrete has been studied by a number of researchers. The range of values found is

(a) elastic conditions, uncracked: 0·01
(b) elastic conditions, cracked: 0·02–0·03
(c) inelastic conditions: 0·03–0·07.

The value of damping has been found to increase with amplitude and increases when the member has been subjected to forces sufficient to cause cracking, so that damping increases for the post-earthquake situation.

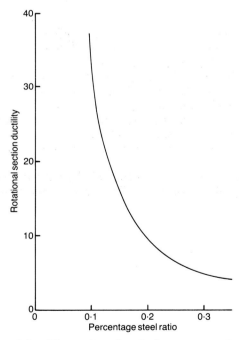

*Fig. 7.27. Rotational ductility and steel ratio for prestressed concrete members*

### 7.11.3. Ductility

The rotational ductility requirement for individual members is from 3 to 5 times the displacement ductility of the frame of which it forms part. The FIP's recommendations propose a minimum rotational ductility value for members of 10. Fig. 7.27 shows a typical relationship between section ductility and steel ratio, from which it can be seen that the recommended minimum corresponds to a steel ratio of 0·002, and this value is recommended as a maximum, unless special calculations are done to show that a higher steel content will still provide the minimum recommended rotational ductility.

Both the FIP recommendations and NZS 3101:1982 advise limiting the depth of the compressive stress block at ultimate load. This ensures adequate ductility in hinge zones. The New Zealand recommendation is to limit it to $0·2h$ and the FIP's to $0·25h$ where $h$ is the overall depth of the section. This requirement will encourage the placing of prestressing steel towards the outer fibres of the section and will discourage central prestressing.

### 7.11.4. Shear

The provision of stirrups follows practice for reinforced concrete. The calculation of shear forces should be based on capacity design principles allowing for overstrength in end moments of 1·15. Transverse reinforcement should be designed to carry the whole of the shear in plastic hinge zones.

## 7.12. Bibliography

Blume, J. A., Newmark, N. M. & Corning, L. H. (1961). *Design of multistory reinforced concrete buildings for earthquake motions.* Chicago: Portland Cement Association.

Martin, L. D. & Korcosz, W. J. (1982). *Connections for precast prestressed concrete buildings.* Prestressed Concrete Institute, Chicago, Technical Report 2.

Park, R. (1986). Ductile design approach for reinforced concrete frames. *Earthquake Spect.* **2**, No. 3, 565–614.

Park, R. & Paulay, T. (1975). *Reinforced concrete structures.* New York: Wiley.

Paulay, T. (1986). The design of ductile reinforced concrete structural walls for earthquake resistance. *Earthquake Spect.* **2**, No. 4, 783–823.

## Chapter 8

# Structural steelwork design

'Structures located in seismic regions must be
designed to resist considerable lateral inertial
loads. The design of such structures requires
a balance between strength, stiffness and
energy dissipation.' Charles Roeder, *J. Struct.
Div. Am. Soc. Civ. Engrs*, 1978, **104**, No. 3, 391.

The scope of this chapter covers

(*a*) earthquake damage—conclusions
(*b*) materials and workmanship
(*c*) analysis and design methods
(*d*) concentric braced frames
(*e*) eccentric braced frames
(*f*) beams
(*g*) columns
(*h*) connections and joint behaviour.

## 8.1. Introduction

Structural steel is in many ways an ideal material for earthquake
resistance. It can exhibit a high level of material ductility and energy
absorption, and experience has shown that steel structures have
usually performed well in earthquakes. However, in order to make
good use of the inherent ductility of steel, considerable care is needed
in the design and detailing of framing systems and connections. The
collapse of the 20 storey Pinot Suarez tower, a steel structure, in the
1985 Mexico earthquake made it clear that designing in steel is not
an automatic passport to survival.

In general, steel structures are more flexible than those of reinforced concrete, and accompanying larger displacements may lead to
higher levels of damage to non-structural components, and to higher
secondary stresses from $P$–$\delta$ effects.

Research has made considerable progress in understanding the differences in performance between monotonic loading and cyclic
loading. Much work has also been done on ductile types of braced
frame, and this work represents a major advance in the conceptual
design of steel structures for earthquakes.

## 8.2. Lessons learned from earthquake damage

Types of damage suffered by structural steelwork are given in

Chapter 1. The conclusions that can be drawn from these are

    (*a*) all elements of the seismic force resisting structure should be designed to be capable of ductile response

    (*b*) all forms of brittle failure, such as bolt fracture in tension or shear, and member buckling must be avoided, even in response to a major earthquake

    (*c*) joints should be provided at discontinuities with adequate provision for movement so that pounding damage cannot occur

    (*d*) non-structural elements should be attached in such a way that they can accommodate displacements that will occur in a major earthquake

    (*e*) failure mechanisms should provide maximum redundancy: the possibility of failure by local collapse, such as would occur if a column failed, should be avoided

    (*f*) all portions of the building should be well tied together.

### 8.3.  Materials and workmanship

Care should be taken in the selection of steel quality. For ductile elements, steel should be low carbon and weldable steel with good notch ductility. Where lamellar tearing is a consideration, a low sulphur content, 0·02% or less, is desirable. The yield strength should not exceed 360 N/mm², and the ratio of ultimate strength to yield strength should exceed 1·4. The upper limit of yield strength should be stipulated, generally not more than 15% greater than the specified value.

Consideration should be given to the possibility of lamellar tearing, a discussion of which is given by McKay (1985). Lamellar tearing can occur where butt or fillet welds of 20 mm or over are made on plates at least 30 mm thick, where there is a high degree of restraint. Tearing can occur in planes parallel to the direction of rolling. The solutions to this problem lie to a limited degree in the selection of the steel (low sulphur content), in inspection procedures, as tearing usually occurs during fabrication following cooling of the adjacent weld, and in good detailing of welds which is discussed in the following section on connections.

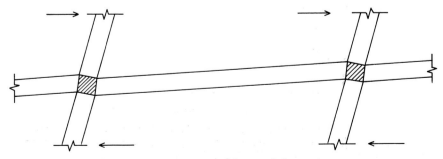

*Fig. 8.1. Frame displacement due to panel zone deformation*

The detailing and fabrication of ductile portions of the structure should consider the possibility of low cycle fatigue—structures responding to earthquakes rarely go through more than 20 cycles of response. Fatigue failure can initiate at notches and cracks which run at right angles to the direction of stress. Welding should follow the best standards of quality and inspection.

Bolt holes should be drilled and not punched or reamed.

For portions of the structure which are designed to remain elastic under major earthquake conditions, normal good practice is required. A comprehensive discussion of quality control and workmanship is given by McKay (1985).

## 8.4. Analysis and design methods

### 8.4.1. Analysis of steel frames

Modelling for analysis requires considerations other than the assumption of members rigidly connected at point nodes. Firstly the buckling strength of compression members needs to be considered. Secondly the effect of distortion of the beam–column panel zone on frame displacement must be taken into account. Fig. 8.1 shows the mechanism of panel zone distortion and Fig. 8.2 shows its relative significance.

In order to give realistic estimates of elastic frame displacements under load it is necessary to model the panel zone as an elastic element. As can be seen from Fig. 8.2 the effect of the panel zone can

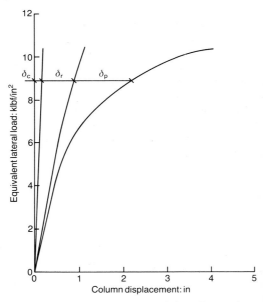

*Fig. 8.2. Column displacement components (after Popov (1980)) ($\delta_c$, displacement due to panel zone deformation; $\delta_r$, displacement due to rotation of the beam; $\delta_p$, displacement due to direct load (P–$\delta$))*

increase markedly as the lateral loading is increased. The effect of the panel zone flexibility can to some extent be compensated for by using stiffener plates to the column.

Secondary stresses in columns, the $P$-$\delta$ effect, may become significant when the structure undergoes large inelastic displacement but are unlikely to matter in the elastic range of response. Increasing the stiffness of a member to deal with this effect is unlikely to be effective: an increase in strength without a significant increase in stiffness is needed.

Where the capacity design approach is used, the design forces in members will be derived in all potentially yielding members, from member strengths, and not directly from the frame analysis. Nevertheless elastic lateral load analysis still forms an essential step in the design process and accurate modelling is still important.

### 8.4.2. Design methods

The approaches to design permitted in codes fall into three principal categories.

The first approach is to derive a set of equivalent lateral forces, and to design the structure on the basis of elastic analysis, but to provide appropriate levels of ductility. Codes world-wide vary on the balance which is struck between strength and ductility, so that it is important that a single consistent approach is used. Patton (1985) gives a range of values of lateral force multipliers and corresponding ductilities for a range of structures, but these apply very specifically to New Zealand practice and should not be used out of context. The New Zealand Ministry of Works and Development (1981) gives lateral force multipliers and appropriate ductilities as shown in Table 8.1.

The second approach is to use plastic design methods. This approach is generally suitable for rigidly connected structures of up to four storeys. Care needs to be taken to ensure that local collapse mechanisms, e.g. in the columns, cannot occur and that displacements remain within acceptable levels.

The third approach is to use the capacity design approach. This is described in Chapter 3, Section 12. Although this method was developed in connection with reinforced concrete structures the principles are quite valid for steel structures. Overstrength factors taking into account material variation and strain hardening may be taken as 1·5 for mild steel and 1·4 for high yield steel.

Table 8.1. *Lateral force multipliers and ductilities*

| Ductility demand | Lateral force multiplier | Ductility |
|---|---|---|
| Full | <2 | >2 |
| Limited | 2–6 | 1–2 |
| None | 6 | 1 |

The normal methods of plastic design should be used to determine the effect of axial load on moment capacity. Unlike concrete sections the effect of moderate axial loads has little effect on the ultimate moment capacity of steel sections.

## 8.5.  Concentric braced frames

The concentric braced frame is commonly in use to resist wind forces but suffers from low system ductility for cyclic loads. The use of rod-type braces which have negligible compression strength is bad practice for seismic structures as there is an undesirable impact-type response with the loss of pre-tension on each cycle.

Because one diagonal of an opposing pair is always in tension, the possibility of a brittle-type failure is present. An additional drawback in the use of X braced panels is that there is effectively no way in which access can be gained through the panel, which places a major restriction on the areas where they can be used.

Code provisions for the design of concentric braced frames for seismic-resistant structures are given by the Structural Engineers Association of California (1985).

## 8.6.  Eccentric braced frames

### 8.6.1.  General

Some types of eccentric braced frames (EBFs) are shown in Fig. 8.3, together with some possible access panels. The primary benefit of the EBF is that it becomes possible to develop substantial system duc-

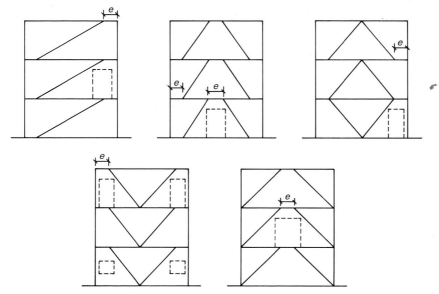

Fig. 8.3. Types of eccentric braced frame (e is the length of the link beam (measured as the clear span))

137

tilities, and the secondary advantage is that, because of the ease with which access can be gained through the plane of the braced panel, they may be located within the building. In design practice they now represent an economically effective way of designing steel structures for earthquake loading.

By selecting a suitable frame stiffness and yield level it is possible to resist moderate earthquakes elastically, with only moderate displacements, and to resist major earthquakes inelastically. One potential drawback is the possibility of floor damage near the link beam during major earthquakes, but in view of the levels of damage normally regarded as acceptable this is not serious.

Sets of design rules for EBFs are given by the Structural Engineers Association of California (1985) and the Building Seismic Safety Council (1985) but these are not reproduced here as they are closely related to other provisions of each code. Nevertheless they provide good practical design guidance.

### 8.6.2. Elastic behaviour

Whether or not a braced frame acts in conjunction with a moment frame, its stiffness is of great importance. Acting with a frame its stiffness will affect the distribution of force between the moment frame and the braced frame. In either case the braced frame will have a major effect on the overall stiffness of the system, thereby determining the level of force that it will be subjected to by moderate earthquakes.

For buildings of three storeys or more, founded on firm material, the seismic displacement response is governed approximately by

$$x_{\text{max}} \approx f(m)k^{0.75} \tag{8.1}$$

where $x_{\text{max}}$ is the maximum displacement, $m$ is the mass and $k$ is the stiffness. Thus it can be seen that the ability to exercise control over stiffness is a valuable tool in the designer's hands.

The elastic design parameters of an EBF can be characterised in the way illustrated in Fig. 8.4 for a simple system. The length assign-

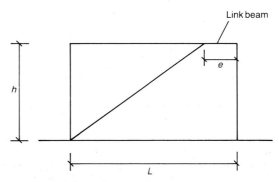

*Fig. 8.4. Eccentric braced frame parameters*

ed to the link, or 'active link', beam is its clear span. The levels over which stiffness can be varied for more complex arrangements can be illustrated with reference to this frame, using the parameter $e/L$ as the control variable. As $e/L$ is varied the system changes from a moment frame with $e/L = 1$ to a concentric braced frame with $e/L = 0$.

Figures 8.5–8.7 show the influence of other frame parameters on the elastic stiffness. Fig. 8.5 varies the frame geometry for a fixed set of section properties. Fig. 8.6 deals with a greater width-to-height ratio where it would be impractical to use a single brace. Fig. 8.7 deals with variations in section properties for a fixed framing geometry. Each relationship shows the very considerable variation in stiffness that is possible, the sensitivity being especially great within the area of most practical configurations with $e/L$ between 0·05 and 0·25.

For $e/L < 0·5$ the shear stiffness of the link beam plays a significant role in the elastic stiffness of the frame. It is clear from the kinematics of the system that shear forces are concentrated in the link beam, and Fig. 8.8 shows the significance of variations in the link beam shear stiffness.

### 8.6.3. Inelastic behaviour

Behaviour in the inelastic range is dominated by the link members, which may have very high ductility requirements. Hjelmstad & Popov

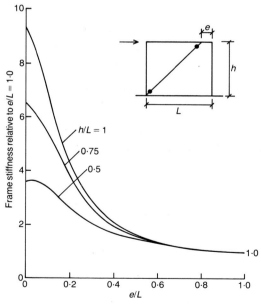

Fig. 8.5. Eccentric braced frame stiffness for varying geometric arrangements (after Hjelmstad & Popov (1984)) ($I_b/I_c = 0·25$, $I_b/A_{br} L^2 = 0·001$, $EI_b/GA_b'L^2 = 0·01$, where $I_b$ is the moment of inertia of the beam, $I_c$ is the moment of inertia of the column, $A_{br}$ is the cross-sectional area of the brace, $A_b'$ is the shear area of the beam, $G$ is the shear modulus and $E$ is the elastic modulus)

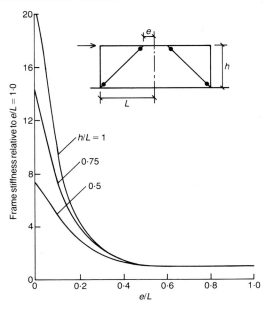

Fig. 8.6. *Eccentric braced frame stiffness for varying geometric arrangements (after Hjelmstad & Popov (1984)*

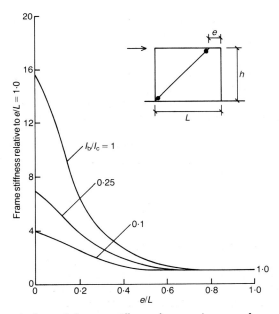

Fig. 8.7. *Eccentric braced frame stiffness for varying member properties (after Hjelmstad & Popov (1984)) ($I_b/A_{br}L^2 = 0.001$, $EI_b/A_b L^2 = 0.01$, $h/L = 0.75$)*

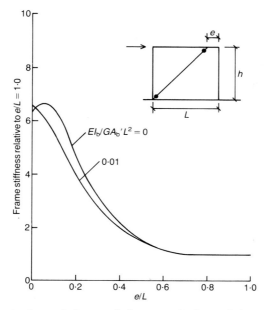

*Fig. 8.8. Eccentric braced frame: influence of shear deformation on frame stiffness (after Hjelmstad & Popov (1984)) ($I_b/A_{br}L^2 = 0.001$, $I_b/I_c = 0.25$, $h/L = 0.75$)*

(1984) estimated values for a three-storey EBF and found a member ductility requirement for the links of 72 which compares with a system ductility of 7·4. This is a very high value.

Hjelmstad & Popov (1984) also noted that the distribution of member forces throughout the structure at ultimate load bore little resemblance to those derived from factored elastic loading. The location of points of contraflexure was completely different: for example columns had gone into single bending, with no reversal of bending moment.

The influence of buckling has been studied but is outside the scope of this work on the grounds that it should be good design practice to ensure that it does not occur.

Estimates of the ductility requirement may be arrived at by elasto-plastic dynamic analysis, but for large structures this is expensive and time consuming. Preliminary estimates may be made from Hjelmstad & Popov (1984) by considering the rigid–plastic deformation of the frame under horizontal load, leading to a relationship between frame and member ductility. The frame ductility demand can be estimated from non-linear spectra or other dynamic considerations.

The admissible deformation shapes for two types of EBF are shown in Fig. 8.9. For the simple frame shown in Fig. 8.10 the relationship between member deformation and structure deformation can be found from

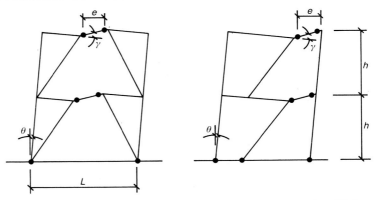

Fig. 8.9. Eccentric braced frames: deformation mechanisms (after Hjelmstad & Popov (1984))

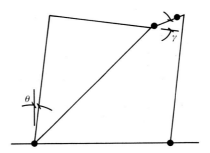

Fig. 8.10. Simple model for an eccentric braced frame collapse mechanism

$$\theta L = \gamma e \tag{8.2}$$

Similar relationships can be derived in the same way for other EBF arrangements.

An alternative approach to the estimation of ductility requirements, also based on rigid–plastic methods, is given by Kasai & Popov (1984).

The inelastic capacity of the shear links themselves depends on the moment shear relationship, as illustrated in Fig. 8.11, the controlling parameters being given by

$$V_p^* = \tau_y(d - t_f)t_w \tag{8.3}$$

$$M_p^* = \sigma_y(b - t_w)(d - t_f)t_f \tag{8.4}$$

$$M_p = \sigma_y Z \tag{8.5}$$

where $\tau_y$ is the yield stress in pure shear, $\sigma_y$ is the yield stress in pure tension, $\sigma_y = \tau_y\sqrt{3}$ for the von Mises yield criterion, $Z$ is the plastic section modulus and $d$, $t_f$, $b$ and $t_w$ are defined in Fig. 8.11.

142

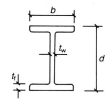

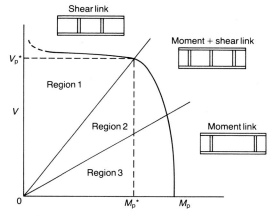

*Fig. 8.11. Moment–shear interaction diagram and appropriate link stiffening details (after Hjelmstad & Popov (1984))*

Hjelmstad & Popov (1984) show that the curve may be approximated by

$$\left(\frac{|M| - M_p^*}{M_p - M_p^*}\right)^2 + \left(\frac{V}{V_p^*}\right)^2 = 1 \qquad M_p^* \leq |M| \leq M_p \qquad (8.6)$$

otherwise $V = V_p^*$ for $|M| \leq M_p^*$

Referring to Fig. 8.11, three kinds of link can be identified. In region 1 where $V < V_p^*$ the link is a shear link and if provided with suitable stiffeners to prevent web buckling will provide excellent ductility under cyclic loading. In regions 2 and 3 the links are moment links, the difference between the two being that in region 2 the shears are still substantial and stiffeners are required for both web shear and flange buckling. In region 3 stiffeners are required for flange buckling only.

The maximum shear hinge length $b^*$ is given by

$$b^* = 2M_p^*/V_p^* \qquad (8.7)$$

Web buckling stiffeners should be placed at $25$–$30t_w$ apart. Some typical details proposed by Hjelmstad & Popov (1984) are shown in Fig. 8.12.

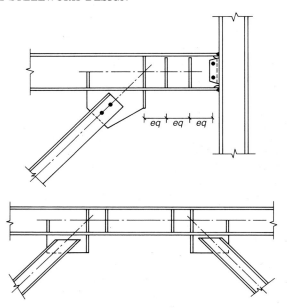

*Fig. 8.12. Typical shear link details (after Hjelmstad & Popov (1984))*

## 8.7.  Design of steel beams

For moment frames it is customary to design on the basis of strong (elastic) columns and weak (inelastic) beams, so that beams are required to provide energy absorption and adequate rotation capacity at certain points. For suitable steels, stable hysteretic yield behaviour can be provided by normal universal beam or wide flange sections as long as buckling can be avoided or controlled.

Lateral or torsional buckling of compression members may lead to sudden collapse and cannot normally be tolerated. However, local buckling of plates is less serious because they retain a substantial post-buckling strength, and this is relevant to beam design as the webs and flanges behave as plates in their buckling performance. Nevertheless buckling will generally tend to 'pinch' the hysteresis loop, reducing the energy absorbed, and this is obviously undesirable. Increased rotations may also result causing distress in some adjacent portion of the structure.

Research has clearly demonstrated that the buckling strength of members subjected to cyclic loading reduces with successive cycles so that the rules adopted in design for cyclic loads need to be considerably more conservative, in comparison with those for statically applied monotonic loads.

Walpole (1985a,b) recommends the following procedure.

(a)  The bending moment diagram is scaled so that the peak value is equal to the plastic moment of resistance of the beam, $M_p$. The length of plastic hinge is then taken as that portion of beam where the moment is greater than $0.85M_p$.

(b) Maximum width-to-thickness ratios are as stated in Table 8.2.

(c) The spacing of lateral restraints is as stated in Table 8.3.

The symbols are explained in Fig. 8.13.

## 8.8. Design of steel columns

For the normal case, columns are designed elastically and follow the normal practice for non-seismically loaded structures. However, cases occur, particularly with ground floor columns, where plastic hingeing can occur in an acceptable way and the following discussion may be applied to such cases.

The rules for lateral buckling are modified to take account of the effect of plastic hinges at each end of a member. Butterworth & Spring (1985), dealing with design for seismic conditions, recommend the placing of flange restraints to the plastic hinge zone. The length of the hinge zone is defined as that length over which the moment exceeds $0.75M_{pc}$, $M_{pc}$ being the plastic moment allowing for the effect of axial compressive load. Over the length $L_y$ flange restraints should be provided at a maximum spacing of $480r_y/f_y^{1/2}$, with a minimum of one restraint being provided within the length. Adjacent restraint should be provided within $720r_y/f_y^{1/2}$.

To combat the effects of local buckling Butterworth & Spring (1985) recommend that the section geometry values given in Table 8.4 should not be exceeded.

The effects of combined load and axial moment may be assessed from the following set of rules given by Butterworth & Spring (1985)

(a) at a support, axial tension or compression

*Table 8.2. Maximum width-to-thickness ratios for fully ductile members*

| Flanges and plates in compression with one unstiffened edge (flanges to beam and channel sections) | $b_1f_y^{1/2}/T$ | 120 |
|---|---|---|
| Flanges of welded box sections in compression | $b_2f_y^{1/2}/T$ | 500 |
| Flanges of rectangular hollow sections | $b_2f_y^{1/2}/T$ | 350 |
| Webs under flexural compression | $d_1f_y^{1/2}/t$ | 1000 |
| Webs under uniform compression | $d_1f_y^{1/2}/t$ | 500 |

*Table 8.3. Spacing of lateral restraints for fully ductile members*

| Flange length where $M > 0.85M_p$ | $> 480r_y/f_y^{1/2}$ | $\leq 480r_y/f_y^{1/2}$ |
|---|---|---|
| Spacing of braces within length where $M > 0.85M_p$ | $\leq 480r_y/f_y^{1/2}$ | One brace required |
| Spacing to brace adjacent to length where $M > 0.85M_p$ | $\leq 720r_y/f_y^{1/2}$ | $\leq 720r_y/f_y^{1/2}$ |

145

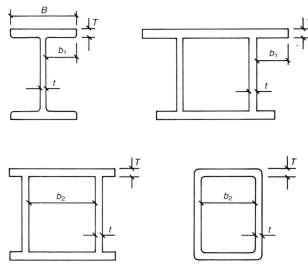

Fig. 8.13. Section nomenclature for Tables 8.2 and 8.3 ($r_y$ is the radius of gyration about the minor axis and $f_y$ $(N/mm^2)$ is the specified yield stress)

Table 8.4. Section geometry limits for fully ductile columns*

| | |
|---|---|
| $b/T$ (WF or UC section flange) | $\leq 120/f_y^{1/2}$ |
| $b/T$ (box flange) | $\leq 500/f_y^{1/2}$ |
| $d/t$ (web) | $\leq 500/f_y^{1/2}$ |

* Notation as in Fig. 8.13.

(i) bending about the major principal axis: for $P/P_y < 0.15$

$$M/M_p \leq 1.0 \qquad (8.8)$$

for $P/P_y \geq 0.15$

$$\frac{P}{P_y} + \frac{M}{1.18 M_p} \leq 1.0 \qquad (8.9)$$

(ii) bending about the minor principal axis: for $P/P_y < 0.4$

$$M/M_p \leq 1.0 \qquad (8.10)$$

for $P/P_y \geq 0.4$

$$\left(\frac{P}{P_y}\right)^2 + \frac{M}{1.18 M_p} \leq 1.0 \qquad (8.11)$$

(iii) bending about both principal axes: for $P/P_y < 0.15$

$$\frac{M_x}{M_{px}} + \frac{M_y}{M_{py}} \leq 1.0 \qquad (8.12)$$

for $P/P_y \geq 0.15$

$$\frac{P}{P_y} + \frac{M_x}{1.18M_{px}} + \frac{M_y}{1.18M_{py}} \leq 1.0 \qquad (8.13)$$

(b) away from a support, axial compression only

(i) bending about the major principal axis: for $P/P_{ac} < 0.15$

$$\frac{P}{P_{ac}} + \frac{M}{M_{px}} \leq 1.0 \qquad (8.14)$$

for $P/P_{ac} \geq 0.15$

$$\frac{P}{P_{ac}} + \frac{C_{mx}}{(1 - P/P_{0cx})M_{0x}} \leq 1.0 \qquad (8.15)$$

(ii) bending about both principal axes: for $P/P_{ac} < 0.15$

$$\frac{P}{P_{ac}} + \frac{M_x}{M_{0x}} + \frac{M_y}{M_{py}} \leq 1.0 \qquad (8.16)$$

for $P/P_{ac} \geq 0.15$

$$\frac{P}{P_{ac}} + \frac{C_{mx}M_x}{(1 - P/P_{0cx})M_{0x}} + \frac{C_{my}M_y}{(1 - P/P_{0cy})M_{py}} \leq 1.0 \qquad (8.17)$$

where $P$ is the axial load in the column, $P_y$ is the squash load of the column $(Af_y)$, $M$ is the column bending moment, $M_p$ is the plastic column moment, zero load, $M_{0x}$ is the maximum column moment capacity, zero load, taking into account lateral buckling and plasticity, $P_{ac}$ is the allowable column load in compression, $P_{0cx}$ is the Euler buckling load $\pi^2 EI_x/l_x^2$, $P_{0cy}$ is the Euler buckling load $\pi^2 EI_y/l_y^2$, $l_x$ and $l_y$ are effective lengths in the bending plane, $M_{py}$ is the plastic moment capacity, zero axial load, in the $y$ direction, and $C_{mx}$ and $C_{my}$ are coefficients used to determine an equivalent uniform bending stress about each axis. Where an elastic buckling analysis is used to determine $M_{0x}$, $C_{mx} = 1$. Otherwise, for members in frames where sidesway is not prevented, $C_m = 0.85$, and for members in frames where sidesway is prevented and not subject to transverse loading between their supports in the plane of bending, $C_m = 0.6 - 0.4\beta$, but not less than $0.4$.

$\beta$ is the ratio of the smaller to the larger moments at the ends of that portion of the unbraced member in the plane of bending under consideration. $\beta$ is positive when the member is bent in reverse curvature and negative when it is bent in single curvature.

## 8.9. Connection design and joint behaviour

### 8.9.1. General

The design of connections in seismically resistant structures differs from non-seismic design because of the necessity to accommodate inelastic response in the members. Although studies of inelastic behaviour in connections have shown that some energy absorption is

147

possible (there is a discussion of this in Nicholas (1985)), normal practice is for connections to be designed to remain elastic.

The forces exerted on connections should be those occurring when there is plastic hingeing in the connecting members with a capacity design approach, and allowing for overstrength and strain hardening. Because of uncertainty in the manner in which a structure will respond to the violent shaking of an earthquake, Nicholas (1985) recommends the following minimum forces for connections

(a) for moment connections, a moment of 1·5 times the connecting member moment based on the nominal yield stress

(b) for non-moment connections one-third of the moment capacity of the connecting member based on the nominal yield stress

(c) for all connections one-half of the strength of the member in tension or compression, based on the nominal yield stress

(d) for all connections 15% of the member strength in shear, based on the nominal shear strength.

### 8.9.2. *Welding*

Normal good practice will apply in welding, but the highest standard is needed because of the possibility of low cycle fatigue. The avoidance of lamellar tearing which was discussed in Section 8.3 is assisted by detailing to avoid cross-plate stressing by welds under conditions of constraint. Fig. 8.14 shows some recommended detailing practice.

Nicholas (1985) gives advice on welding which includes the following points. The best form of load transfer is a full penetration butt weld, where the weld material strength is greater but not significantly greater than the parent metal. Partial penetration butt welds should not be used where cyclic stressing will occur. Fillet welds are acceptable if the following rules are observed.

(a) Intermittent welding should be minimised as the ends of runs are stress raising discontinuities.

(b) The throat thickness should not be less than half the plate thickness.

(c) Tearing stresses in the parent metal should be checked where high strength electrodes are used, and the leg length of the weld is small.

### 8.9.3. *Beam and column splice joints*

Beam and column splice joints will normally be designed in a similar way to those used in non-seismic design. Their location should generally be in zones of low stress, and as long as the loading criteria given in Section 8.9.1 are accommodated their design will be straightforward. In the unusual event that they are located in a potential plastic hinge zone, special consideration will need to be given.

### 8.9.4. *Beam–column joints and the panel zone*

Some types of beam–column joint are shown in Figs 8.15–8.18. The

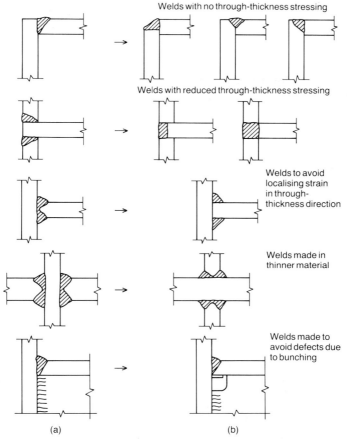

Welds with no through-thickness stressing

Welds with reduced through-thickness stressing

Welds to avoid localising strain in through-thickness direction

Welds made in thinner material

Welds made to avoid defects due to bunching

(a)                    (b)

*Fig. 8.14. Weld details to avoid lamellar tearing (after McKay (1985)): (a) not recommended; (b) recommended*

choice of type will depend more on economics, the available skills, quality control and fabrication resources than on design requirements. Each of the types shown can function as a moment connection. The all-welded types shown in Figs 8.15 and 8.16 can be provided with bolted web shear connectors in place of the welded web shown. This (bolted web) type has been satisfactorily tested in the laboratory.

Where connections are bolted to the column as in Figs 8.17 and 8.18 care needs to be taken over the effects of bending distortion on the end plate or head of the Tee connector. Fig. 8.19 shows the effect of prying on the Tee connection, and a similar effect may occur with the welded end plate. Recommendations for design are made by Douty & McGuire (1965). For normal sections the value of $Q$, the prying force, does not increase the bolt force by more than 10% and only exceeds this where the Tee flange is unusually thin or the bolts are spaced

149

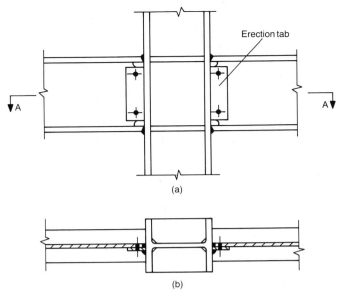

*Fig. 8.15. Welded beam–column connection: (a) elevation; (b) plan on AA*

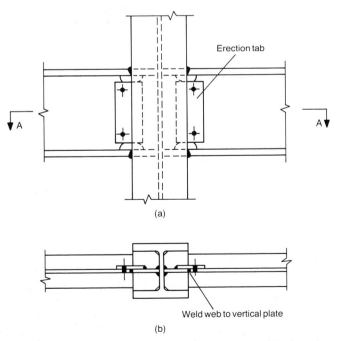

*Fig. 8.16. Welded beam–column connection—weak axis: (a) elevation; (b) plan on AA*

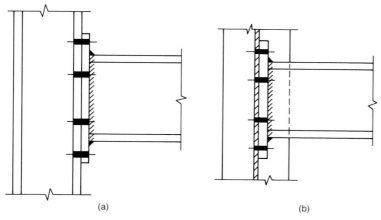

*Fig. 8.17. Bolted beam–column connections: (a) strong axis; (b) weak axis*

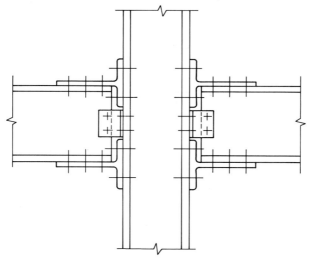

*Fig. 8.18. Bolted beam–column connection using Tees and high strength friction grip bolts*

further from the root than necessary. A design method is given by the American Institute of Steel Construction (1984).

Types of stiffener used in the panel zone are shown in Fig. 8.20. The behaviour of the flange will depend on the support which it derives from stiffeners and is normally analysed on the basis of yield line theory. A summary of standard cases that have been studied is given by Walpole (1985a,b). Where stiffening is required it can be provided by stiffening plates of the general type shown in Fig. 8.20, or reinforcing plates welded directly to the flange as shown in Fig. 8.21.

151

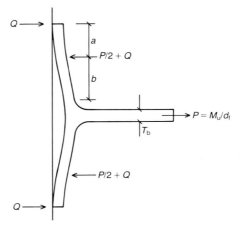

*Fig. 8.19. Effect of prying on bolt forces in Tee connectors*

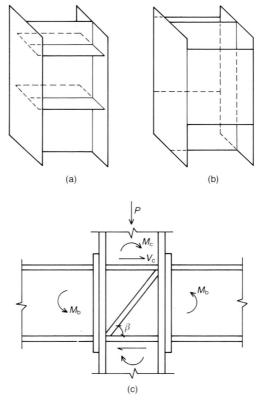

*Fig. 8.20. Column stiffeners in the panel zone: (a) web stiffeners; (b) doubler plates; (c) diagonal stiffener*

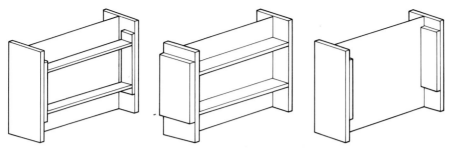

*Fig. 8.21. Reinforcing plates to flanges in the panel zone*

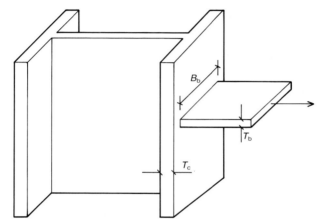

*Fig. 8.22. Application of beam flange load*

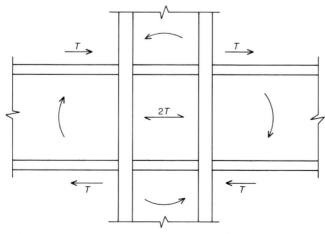

*Fig. 8.23. Shear forces in the panel zone*

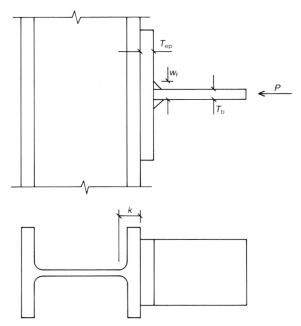

*Fig. 8.24. Parameters affecting load dispersion from the beam flange ($f_{yb}$ is the beam yield stress and $f_{yc}$ is the column yield stress)*

Following Walpole (1985a,b) the flange may be left unstiffened where the beam flange load is applied as in Fig. 8.22 if

$$T_c \geq 0 \cdot 6 (B_b \, T_b \, f_{yb}/f_{yc})^{1/2} \tag{8.18}$$

where the symbols are defined in Fig. 8.22.

Equation (8.18) incorporates an allowance for a beam overstrength factor of 1·5.

Degenkolb (1978) illustrates the shear forces acting on the panel zone due to lateral loading. It can be seen from Fig. 8.23 that the panel zone shear is $2T$ and, a point which is often not realised by designers, the total shear taken by the weld between the stiffener and web is also $2T$.

Witteveen, Stark, Bijlaard & Zoetemeijer (1982) give the following limits for the unstiffened web, the factor of 1·5 being the beam over-strength factor

$$t_c \geq \frac{1 \cdot 5 B_b \, T_b}{T_b + 5k + 2T_{ep} + 2w_f} \frac{f_{yb}}{f_{yc}} \tag{8.19}$$

where the symbols are defined in Fig. 8.24.

### 8.10. Bibliography

American Institute of Steel Construction (1978). *Specification for the design, fabrication and erection of structural steel for buildings.* Chicago: AISC.

Building Seismic Safety Council (1985). *NEHRP recommended provisions for the development of seismic regulations for new buildings.* Washington DC: Building Seismic Safety Council.

Structural Engineers Association of California (1985). *Tentative lateral force requirements.* Sacramento: SEAC.

Study Group for the Seismic Design of Steel Structures (1985). *Bull. N.Z. Nat. Soc. Earthquake Engng,* **18**, Dec., No. 4, Sections A–K.

# Chapter 9

# Foundations

'There is no glory in foundations.'
Professor K. Terzaghi

The scope of this chapter covers

(a) soil deformation
(b) soil properties
(c) dynamic soil–structure interaction
(d) bearing foundations
(e) piled foundations
(f) retaining walls.

## 9.1. Design objectives

Building structures will not normally be required to deal with massive permanent soil movements such as surface faulting and landslides. However, the effects of possible earthquake-induced settlement or liquefaction may have to be considered.

The design of foundations for earthquake resistance differs from non-seismic design in the following features

(a) the interaction of building and ground dynamics
(b) the nature of the forces transmitted, in particular yield level forces from the structure, and the much more frequent occurrence of uplift forces
(c) the need to provide for dynamic soil distortion.

A horizontal shear is imparted to the building by movement of the supporting ground and it is customary to make design provision for this. Nevertheless, except for small rigid buildings, shear failure between the foundations and ground in earthquakes is uncommon.

The provision of ductility in foundations is unusual, and they would normally be designed to remain elastic. Although there may be some merit in including the foundation in the energy-absorbing ductile system, the difficulties of repair and the possible corrosion of cracked buried members usually make this unacceptable.

## 9.2. Soil deformation
### 9.2.1. Settlement

Dry sands may consolidate as a result of shaking. Generally they will reach a stable density after about 30 s so that it is reasonable to assume that this state will be reached in an earthquake. Newmark &

Rosenblueth (1971) suggest a simple approximate method, based on the assumption that there is a stable void ratio $e_f$ below which level no consolidation will occur. $e_f$ is related to the maximum and minimum practicable void ratios $e_{max}$ and $e_{min}$ respectively by

$$e_f = e_{min} + (e_{max} - e_{min}) \exp(0{\cdot}076x_g) \qquad (9.1)$$

where $x_g$ is the maximum ground displacement in metres.

If the initial void ratio is not greater than $e_f$ no consolidation will occur. The maximum settlement for a soil with a void ratio $e$ will then be

$$s = \frac{(e - e_f)H}{1 - e} \qquad (9.2)$$

where $H$ is the depth of soil. It should be emphasised that this represents a maximum value and is an approximation based on low confining pressures and dry soil.

### 9.2.2. Liquefaction

When a saturated granular soil is shaken over a period, the pore-water pressure will tend to increase. When this pressure reaches the confining pressure the soil will suffer a sharp drop in strength, its behaviour becoming close to that of a liquid. There are numerous examples of this in earthquakes and the resulting damage can be spectacular. In Niigata City, Japan, in 1964 the damage from liquefaction included complete failure of bearing foundations of multi-storey buildings, causing them to topple.

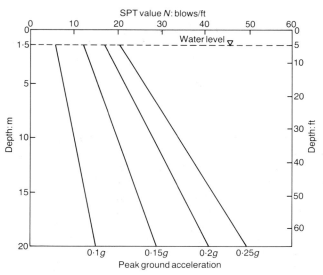

Fig. 9.1. *Standard penetration test values for which liquefaction is unlikely to occur (after Seed & Idriss (1971)) (liquefaction is highly unlikely for any soil if SPT values exceed those shown)*

158

The first assessment of the liquefaction potential may be based on standard penetration tests (SPTs), which will normally be carried out in any site investigation as a method of estimating in situ densities of cohesionless soils. Fig. 9.1 shows boundary SPT values plotted against depth, for a condition with a water-table depth of 1·5 m (5 ft) and a level surface. These have been derived from field studies of earthquake sites where liquefaction has occurred. It should be remembered in comparing field results with these tables that liquefaction does not originate at the surface: at Niigata City the liquefaction layer was in the zone between 4·5 m and 8·5 m deep.

Figure 9.2 is similar to Fig. 9.1 but shows zones of uncertainty for four different cases.

The principal factors which reduce the susceptibility of a soil to liquefaction are

(a) an increase in density
(b) a decrease in the duration of shaking
(c) an increase in the confining pressure
(d) an increase in the grain structure stability
(e) an increase in the time that the soil has been under sustained pressure
(f) an increase in the overconsolidation ratio.

An alternative approach, described by Seed (1979), calculates the cyclic stress ratio (the ratio of cyclic shear stress on a horizontal surface caused by the earthquake to the vertical effective stress) and compares this with either field records or laboratory data. Details of this method are outside the scope of this text but it is a method recommended by the Building Seismic Safety Council (1985) in terms which make it suitable for use by specialist soils engineers. Estimates of liquefaction potential based on laboratory testing of samples should be treated with reserve as it is extremely difficult to reproduce both the sample in its original condition and the conditions of restraint under which it existed.

## 9.3. Soil properties and testing
### 9.3.1. General

Soils behave non-linearly so that the assignment of elastic properties is necessarily an equivalent linearisation at a particular strain level. Field testing will generally be carried out at a very much lower strain level than can be anticipated in strong ground motion from earthquakes. Laboratory tests will be at a relatively high strain and may in some cases reach earthquake levels.

Field testing will generally be able to establish the shear wave velocity $v_s$ and the compression wave velocity $v_p$, measured at low strain. Laboratory tests can make direct measurements on samples for Young's modulus $E$ and the shear modulus $G$ at high strains.

Silver (1981) lists the available procedures for obtaining dynamic soil data as shown in Table 9.1.

The shear wave velocity $v_s$ and the compression wave velocity $v_p$

159

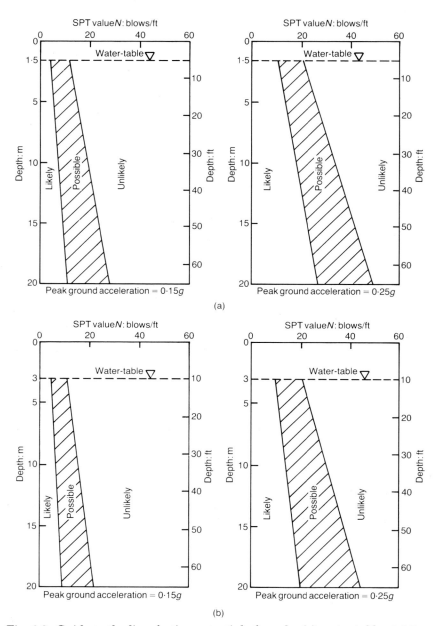

Fig. 9.2. *Guide to the liquefaction potential of sands:* (a) *water-table at 1·5 m;* (b) *water-table at 3 m (after Seed & Idriss (1971))*

*Table 9.1.  Procedures for obtaining dynamic soil data*

| Property | Method* |
|---|---|
| Gradation and soil classification | E, laboratory testing |
| | G, cross-hole and uphole/downhole surveys; reflection† |
| Degree of saturation | E, laboratory testing |
| | G, lateral resistivity |
| Density and relative density | E, laboratory testing |
| | G, cross-hole and uphole/downhole surveys; reflection† |
| | I, SPT |
| Dynamic modulus | E, laboratory testing |
| | G, cross-hole and uphole/downhole surveys; reflection |
| | I, SPT |
| Damping | E, laboratory testing |
| | G, in situ impulse |
| Dynamic strength | E, laboratory testing |
| | G, no procedure available |
| | I, SPT |

* E represents conventional procedures; G represents geophysical procedures; I represents conventional in situ procedures.
† Data obtained by these procedures may be based on correlations with such factors as P wave velocities, S wave velocities, shear modulus, Young's modulus and Poisson's ratio.

are related to the maximum values of Young's modulus $E_{max}$ and shear modulus $G_{max}$ by

$$G_{max} = \rho v_s{}^2 \tag{9.3}$$

$$E_{max} = \rho v_p{}^2 \tag{9.4}$$

where $\rho$ is the density of the soil and Système International units are used. Also $E_{max}$ and $G_{max}$ may be related to Poisson's ratio by

$$E_{max} = 2(1 + v)G_{max} \tag{9.5}$$

### 9.3.2.  Poisson's ratio

Typical values of Poisson's ratio are given in Table 9.2. Poisson's ratio may also be obtained from field measurements of the compression and shear wave velocities, using the relationship

$$v = \frac{(v_p/v_s)^2 - 2}{2(v_p/v_s)^2 - 2} \tag{9.6}$$

### 9.3.3.  Shear modulus

For sands and cohesive soils the dynamic modulus may be estimated from formulae given by Hardin & Drnevich (1972a,b). For sands guidance may be more simply obtained from Fig. 9.3.

An indication of the variation in the shear modulus with strain is given in Fig. 9.4.

161

Table 9.2. Typical values of Poisson's ratio

| | |
|---|---|
| Saturated clay | 0·4–0·5 |
| Unsaturated clay | 0·1–0·3 |
| Sandy clay | 0·2–0·3 |
| Silt | 0·3–0·35 |
| Dense sand | 0·2–0·4 |
| Coarse sand | 0·15 |
| Fine sand | 0·25 |

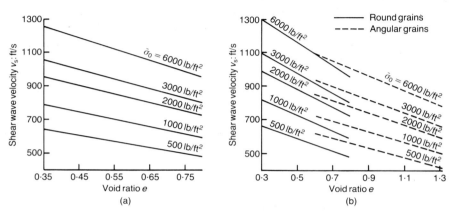

Fig. 9.3. Shear wave velocities for sands (after Hardin & Richart (1963)): (a) variation in shear wave velocity with void ratio for dry Ottawa sand; (b) variation in shear wave velocity with void ratio for dry round and angular sands

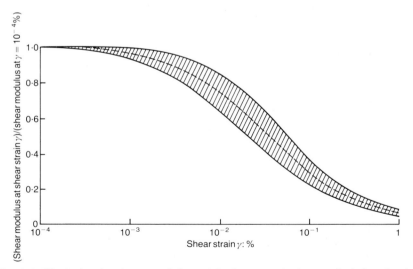

Fig. 9.4. Variation in shear modulus with shear strain for sands (after Seed & Idriss (1970))

For cohesive soils values of $G$ will normally be obtained from the testing of samples. However, for preliminary assessment Fig. 9.5 can be used when the undrained shear strength $S_u$ is known or can be estimated.

### 9.3.4. Young's modulus
Young's modulus can usually be derived from tests, but some guidance on typical values may be taken from Table 9.3.

### 9.3.5. Damping
The hysteretic damping ratio for soil is not dependent on density and may be derived from the maximum damping value at large strains, for both cohesive and non-cohesive soils from (Hardin & Drnevich, 1972a, b)

$$\xi = \frac{\xi_{max}\,\gamma}{\gamma_r + \gamma} \tag{9.7}$$

Table 9.3. *Typical Young's modulus values for soils*

| Soil type | $E$: N/mm$^2$ |
|---|---|
| Soft clay | Up to 15 |
| Firm to stiff clay | 10–50 |
| Very stiff, hard clay | 25–200 |
| Silty sand | 7–70 |
| Loose sand | 15–50 |
| Dense sand | 50–120 |
| Dense sand and gravel | 90–200 |

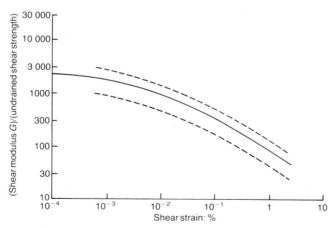

Fig. 9.5. *Normalised shear modulus for saturated clays (after Seed & Idriss (1970))*

163

where $\xi$ is the viscous damping ratio, $\xi_{max}$ is the maximum damping ratio at large strain, $\gamma$ is the strain and $\gamma_r$ (maximum shear strain/$G_{max}$) is the reference strain.

The maximum damping ratio for sands, $\xi_{max}$, is given by

$$\xi_{max} = \frac{D - 1\cdot 5 \log N}{100} \qquad (9.8)$$

where $D = 33\%$ for clean dry sands, $D = 28\%$ for clean saturated sands and $N$ is the number of cycles. The inclusion of $N$ in equation (9.8) is explained by Hardin & Drnevich (1972a,b), who show that small changes in the damping value occur with the number of effective cycles to which the soil is subjected.

For cohesive soils

$$\xi_{max} = 31 - (3 + 0\cdot 3f)\bar{\sigma}_m^{1/2} + 1\cdot 5f^{1/2} - 1\cdot 5 \log N \qquad (9.9)$$

where $f$ is the frequency of the applied load in hertz and $\bar{\sigma}_m$ is the mean principal effective stress in kilograms per centimetre squared. Figs 9.6 and 9.7 show some typical values.

### 9.4. Soil–structure interaction

The dynamic basis of soil–structure interaction is discussed in Chapter 3. The approach dealt with here is where the soil underlying the building is sufficiently elastic for flexibility in the soil beneath the foundations to affect the dynamic behaviour of the structure. Generally the effect of introducing flexibility at foundation level will reduce the base shear and forces in members. However, because a rocking mode is introduced, displacements may be increased, which can have two important consequences. Firstly secondary stresses due to $P$–$\delta$ effects in the columns may become significant. Secondly, the

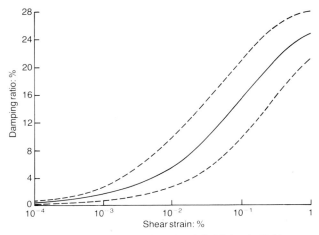

*Fig. 9.6. Damping ratios for sands (after Seed & Idriss (1970))*

increased accelerations and displacements in the building will adversely affect the contents and architectural components.

Regulations for the design of nuclear structures provide for uncertainty in soil–structure interaction effects by considering variations in the shear modulus $G$ from the design value used. This should be based on engineering judgement, considering the accuracy of estimation, but a minimum spread of values is defined as ranging from $0.67G$ to $1.5G$.

The type of model used to simulate soil flexibility under a footing is illustrated in Fig. 9.8.

Damping at the soil–foundation interface arises from two sources. Firstly there is material damping arising from the non-linear properties of the soil. Secondly there is radiation damping due to the transmission of energy away from the interface by radiating waves. The damping values given in the following text incorporate damping from both sources.

Equivalent linear springs and viscous damping values are given in Tables 9.4 and 9.5 and Fig. 9.9, based on Whitman & Richart (1967).

Alternative formulae are given by the Building Seismic Safety Council (1985) for vertical and rocking springs, who also provide approximate formulae to allow for the effects of limited depth of stratum and of embedment.

## 9.5.  Bearing foundations

The design of bearing foundations for earthquake-resistant structures does not depart substantially from non-seismic design. Some consideration must be given to the question of how earthquake forces are transmitted from the ground to the structure. The two available

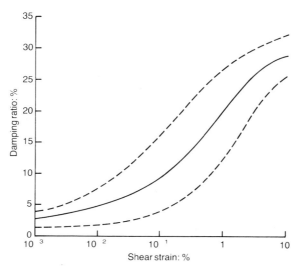

*Fig. 9.7. Damping ratios for saturated clays (after Seed & Idriss (1970))*

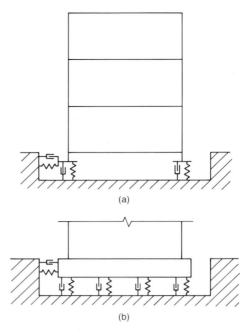

(a)

(b)

*Fig. 9.8. Modelling of springs and dampers under foundations: (a) separate footings with tie; (b) raft foundation*

*Table 9.4. Foundation interaction: spring and damper values for circular bases\**

| Motion | Equivalent spring constant | Equivalent damping coefficient |
|---|---|---|
| Horizontal | $k_x = \dfrac{32(1 - v)GR}{7 - 8v}$ | $c_x = 0{\cdot}576 k_x R(\rho/G)^{1/2}$ |
| Rocking | $k_\psi = \dfrac{8GR^3}{3(1 - v)}$ | $c_\psi = \dfrac{0{\cdot}30}{1 + B_\psi} k_\psi R(\rho/G)^{1/2}$ |
| Vertical | $k_z = \dfrac{4GR}{1 - v}$ | $c_z = 0{\cdot}85 k_z R(\rho/G)^{1/2}$ |
| Torsion | $k_t = 16GR^3/3$ | $c_t = \dfrac{(k_t I_t)^{1/2}}{1 + 2I_t/\rho R^5}$ |

\* $v$ is Poisson's ratio for the foundation medium, $G$ is the shear modulus for the foundation medium, $R$ is the radius of the circular basemat, $\rho$ is the mass density of the foundation medium, $B_\psi = 3(1 - v)I_0/8\rho R^5$, $I_0$ is the total mass moment of inertia of the structure and basemat about the rocking axis at the base and $I_t$ is the polar mass moment of inertia of the structure and basemat.

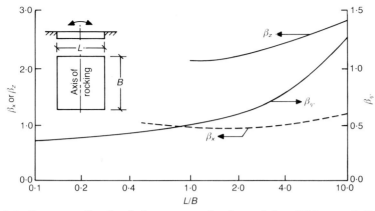

*Fig. 9.9. Constants $\beta_x$, $\beta_\psi$, $\beta_z$ for rectangular bases (after Whitman & Richart (1967))*

*Table 9.5. Foundation interaction: spring and damper values for rectangular bases\**

| Motion | Equivalent spring constant | Equivalent damping coefficient |
|---|---|---|
| Horizontal | $k_x = 2(1 + v)G\beta_x(BL)^{1/2}$ | Use the results for a circular base with the following equivalent radius $R$ |
| Rocking | $k_\psi = \dfrac{G}{1 - v} \beta_\psi \, BL^2$ | $R = (BL/\pi)^{1/2}$ for translation $R = (BL^3/3\pi)^{1/4}$ for rocking |
| Vertical | $k_z = \dfrac{G}{1 - v} \beta_z(BL)^{1/2}$ | |
| Torsion | Use Table 9.4 for $R = [BL(B^2 + L^2)/6\pi]^{1/4}$ | |

\* $v$ and $G$ are as defined in Table 9.4, $B$ is the width of the basemat perpendicular to the direction of horizontal excitation, $L$ is the length of the basemat in the direction of horizontal excitation and $\beta_x$, $\beta_\psi$ and $\beta_z$ are constants that are functions of the dimensional ratio $L/B$ (see Fig. 9.9).

mechanisms in the soil are friction at the interface with the foundation and active pressure at the vertical faces.

The development of lateral pressures on the foundation sides is difficult to justify, bearing in mind that it is the ground that moves the building and not the other way round, and is not permitted in some codes. However, where friction is insufficient to withstand the lateral forces involved it is difficult to understand how else it can be dealt with. In view of the fact that failures in sliding are almost unknown it is difficult to justify expensive measures like piling solely to deal with this problem.

The tying together of individual footings and pads is undoubtedly good practice, as it ensures that differential lateral movements

167

between them are resisted. A common approach is to design these ties to resist 10% of the vertical load on any column to which it attaches, either in tension or compression. It is not essential that the ties are provided at foundation level, and it is common to provide ties that are integral with the ground floor.

Where capacity design principles have been used for the super-structure design, the same concept should be carried into the foundation design. Taylor & Williams (1979) examine the implications of this, including the effects of possible uplift and local yielding. When conditions make such effects unacceptable, the loading on the foundation must be taken as the overstrength yielding condition of the supported structure.

### 9.6.  Piled foundations

Piles have little effect on the dynamic distortion of the surrounding soil and can generally be assumed to comply with this distortion, including possible failure in shear when the surrounding soil distorts. The commonest point of failure in piles is at their junction with the pile cap, and this is more likely if there is a short length of pile beneath the pile cap which is not laterally supported by the ground. This type of failure is less likely when the pile foundation is well keyed in to the surrounding ground by pile caps, tie-beams and deep basements.

Batter piles do not perform well under earthquake loading and are particularly likely to fail at the junction with the pile cap. In general

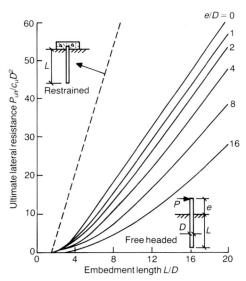

*Fig. 9.10. Ultimate lateral pile resistance for cohesive soils related to embedment length (after Broms (1965))*

they should not be used and the lateral component of thrust taken by vertical piles.

The dynamic effect of piles is that they have little effect on the lateral stiffness of the ground in which they are placed. However, they have a considerable effect on the rocking mode of the building, increasing the stiffness substantially.

The response of piles is considerably affected by liquefaction. Large movements may occur where lateral support is temporarily missing and high bending moments can occur in the pile.

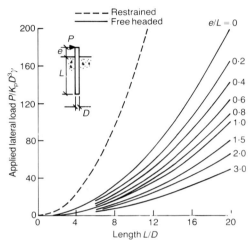

*Fig. 9.11. Ultimate lateral pile resistance for cohesionless soils related to embedment length (after Broms (1965))*

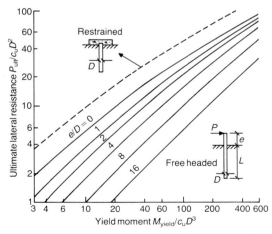

*Fig. 9.12. Ultimate lateral pile resistance for cohesive soils related to yield moment (after Broms (1965))*

169

In complying with lateral movements of the ground, piles will develop curvatures which will tend to be sharpest at the transition boundaries between strata, especially where the soil properties vary substantially. Because moments are developed from curvature imposed from the movement of the ground, stresses are proportional to Young's modulus $E$ and to the pile thickness, so that there is a reason for using the thinnest possible section and pile material with a low stiffness. A failure of piles in the lower sections is unlikely.

Code requirements for piles in seismic zones typically do not permit unreinforced piles and require high ductility in the upper levels, where yielding may occur in major earthquakes.

The resistance of piles to lateral loading may be estimated from Broms (1965), bearing in mind that this approach is developed for static lateral forces. Figs 9.10–9.13 show appropriate values for the ultimate load, where $c_u$ is the undrained shear strength, $K_p$ is the Rankine lateral pressure coefficient and $\gamma$ is the soil's density.

On the basis of the approach used by Broms, bending moments in piles can also be estimated, and these are illustrated in Figs 9.14–9.16.

Once a pile has developed a rotational hinge the distribution of soil pressure changes significantly. Hinges will tend to form in long rather than short piles. It is assumed that for a depth of 1·5 times the pile diameter cohesive soil does not develop significant lateral pressure. Fig. 9.14 shows the distribution for unrestrained piles in cohesive soil, for the yielded and unyielded condition, where $H_u$ is the ultimate lateral load and $f$ is the depth to the maximum bending moment. In this case

$$f = \frac{H_u}{9c_u d} \qquad (9.10)$$

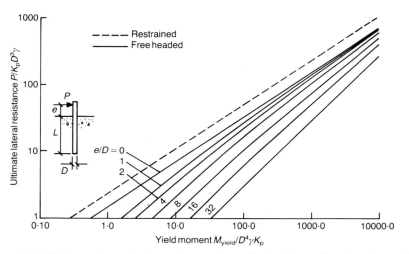

*Fig. 9.13. Ultimate lateral pile resistance for cohesionless soils related to yield moment (after Broms (1965))*

Figure 9.15 shows the conditions applying for short, intermediate and long piles in cohesive soil. For the short pile

$$M_{max} = H_u(0{\cdot}5L + 0{\cdot}75d) \tag{9.11}$$

Yielding of the pile will first occur at the head. Values for $f$, $g$ and $M_{max}$ can be derived from equating the horizontal forces and resolving the moment about the pile head. When $M_{max}$ reaches $M_u$ the distribution of forces and moments will become that shown for long piles.

For piles in non-cohesive soils the conservative assumption is made of an ultimate soil resistance that is three times the Rankine passive pressure, or

$$P_u = 3\sigma_v'K_p \tag{9.12}$$

where $\sigma_v'$ is the effective vertical overburden pressure, $K_p = (1 + \sin \phi')(1 - \sin \phi')$ and $\phi'$ is the angle of internal friction (effective stress).

Figure 9.16 shows the distribution of forces in short and long unrestrained piles. For the unyielded pile the bending moment diagram is derived directly from the distribution of soil pressure. Yielding will occur at the depth $f$ corresponding to the maximum moment and the conditions shown for the long pile will then apply.

The conditions for restrained piles in cohesionless soil are shown in Fig. 9.15. These only differ from those shown for cohesive soil in the triangular soil pressure distribution and the development of a force $F$ at the pile tip in the intermediate length condition, where

$$F = 1{\cdot}5\gamma dL^2K_p - H_u \tag{9.13}$$

For more accurate assessments of lateral force and bending moments, one of the commercially available computer programs which evaluates the seismic response of a multilayered soil should be used with pile moments based on full compliance with the soil.

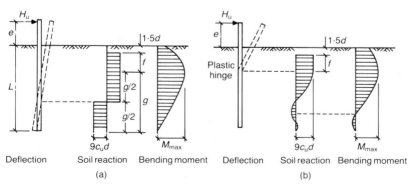

*Fig. 9.14. Static lateral load on unrestrained piles in cohesive soil (after Broms (1965)): (a) short pile; (b) long pile*

171

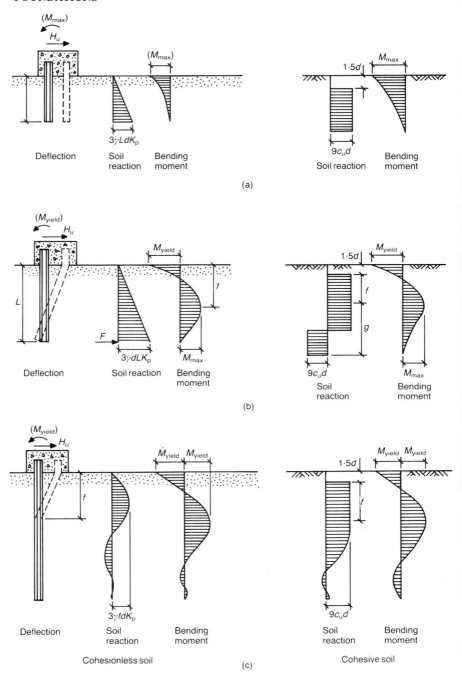

Fig. 9.15. *Static lateral load on restrained piles (after Broms (1965)): (a) short pile; (b) intermediate pile; (c) long pile*

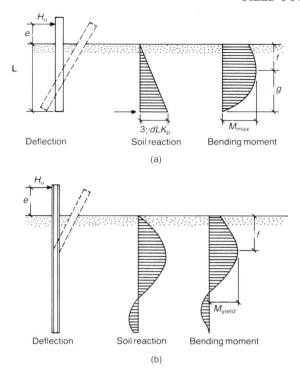

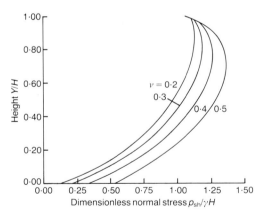

*Fig. 9.16. Static lateral load on unrestrained piles in cohesionless soil (after Broms (1965)): (a) short pile; (b) long pile*

*Fig. 9.17. Dynamic soil pressures for the elastic solution (H is the embedment height, Y is the distance above the base of the retaining structure, $\gamma$ is the unit weight of the retained soil, $v$ is Poisson's ratio and $p_{sh}$ is the lateral dynamic soil pressure for 1·0g peak horizontal earthquake acceleration)*

173

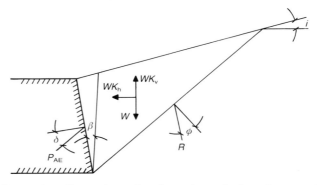

*Fig. 9.18. Dynamic soil pressures for the active solution: forces considered in the Mononobe–Okabe analysis ($p_s = \frac{1}{2}\gamma H^2(1 - k_v)K_{AE}$ where*

$$K_{AE} = \cos^2(\phi - \theta - \beta)[\cos\theta\cos^2\beta\cos(\delta + \beta + \theta)]^{-1}$$

$$\left[1 + \left(\frac{\sin(\phi + \delta)\sin(\phi - \theta - i)}{\cos(\delta + \beta + \theta)\cos(i - \beta)}\right)^{1/2}\right]^{-2}$$

*$\theta = \tan^{-1}[k_h/(1 - k_v)]$, $\gamma$ is the unit weight of the soil, $H$ is the height of the wall, $\phi$ is the angle of friction of the soil, $\delta$ is the angle of wall friction, $i$ is the slope of the ground surface behind the wall, $\beta$ is the slope of the back of the wall from vertical, $k_h(g)$ is the horizontal ground acceleration, $k_v(g)$ is the vertical ground acceleration and $w$ is the weight of the wedge)*

## 9.7. Retaining walls

Solutions to the earth pressure acting on a retaining wall during an earthquake depend on the amount of movement which is acceptable. The independent type of wall, gravity or cantilever, may be able to move a substantial amount without loss of function, and in this case an 'active' solution to the earth pressure is applicable. For basement walls, any movement of consequence will involve a loss of function and in this case an elastic solution will apply.

A common sense bound on the total earth pressure acting on any wall, seismic plus static pressure, is that it should not exceed the calculated passive pressure.

For the elastic solution $p_{sh}$, the dynamic earthquake pressure can be found from Fig. 9.17. The resultant force should be taken to act at 0·6H from the base. This method is conservative for softer soils. The total force on the wall is the sum of the dynamic pressures calculated by this method and the static pressures calculated in the normal way.

For more complex situations, in particular those where other structures interact with the ground supported by the wall, a finite element soil model is necessary. Forces exerted by the interacting structures should include the full seismic base shears and moments.

For the active solution, the Mononobe–Okabe method is commonly used. This solution was derived for dry, cohesionless soils but provides reasonable values for cohesive soils as well, and is based on the

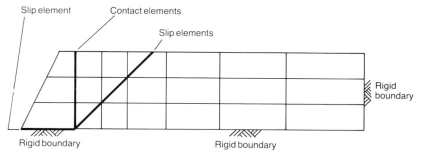

*Fig. 9.19. Finite element model for the active solution incorporating the failure plane*

assumption that sufficient yield develops in the wall to produce the conditions shown in Fig. 9.18, with the maximum shear developed along the sliding wedge surface. The resultant pressure $P_s$ acts at $2H/3$ from the base and variations in pressure up the face of the wall can be taken as a triangle with a maximum pressure at the surface.

The Mononobe–Okabe method has been developed further by Richards & Elms (1979) and Whitman & Liao (1984), but, considering the basic simplicity and the approximating assumptions inherent in the original formulation, it is reasonable either to use the original approach or to use a finite element method incorporating slip elements along the shear plane, as shown in Fig. 9.19. Tables for the calculation of active earth pressures are given by the New Zealand Ministry of Works and Development (1973).

Sherif & Fang (1984) have extended the active solution to a rigid wall rotating at the base and provide design graphs for sand backfill.

## 9.8. Bibliography

Hardin, B. O. (1978). The nature of stress–strain behaviour for soils. *Proc. Specialty Conf. Earthquake Engineering and Soil Dynamics, Pasadena*, 3–90. New York: American Society of Civil Engineers.

Newmark, N. M. & Rosenblueth, E. (1971). *Fundamentals of earthquake engineering*. Englewood Cliffs: Prentice-Hall.

New Zealand Ministry of Works and Development (1973). *Retaining wall design notes*. Civil Division Publication CDP 702/C. Wellington: NZMWD.

Poulos, H. G. & Davis, E. M. (1980). *Pile foundation analysis and design*. New York: Wiley.

Seed, H. B. & Idriss, I. M. (1982). *Ground motions and soil liquefaction during earthquakes*. Berkeley: Earthquake Engineering Research Institute.

Wolf, J. P. (1985). *Dynamic soil–structure interaction*. Englewood Cliffs: Prentice-Hall.

Woods, R. D. (1978). Measurement of dynamic soil properties, *Proc. Specialty Conf. Earthquake Engineering and Soil Dynamics, Pasadena*, 91–180. New York: American Society of Civil Engineers.

# Chapter 10

# Masonry

'Damage to unreinforced brick walls was
severe, just as it had been in all previous
earthquakes.' Karl V. Steinbrugge,
Kern County earthquake 1952

The scope of this chapter covers

(a) masonry behaviour
(b) masonry strength
(c) unreinforced masonry in existing buildings
(d) infill masonry interaction with structural frames
(e) reinforced structural masonry.

## 10.1. Design objectives

For unreinforced masonry, which is unsuitable for use in seismic
areas, concern is limited to the study of existing structures which
may be subjected to earthquake forces. Masonry used as infill to the
panels of a structural frame is subjected to forces from the displacement of the frame as well as inertial forces. Additionally the designer
is concerned with the interaction between the masonry and the frame
which may modify the frame's response and the forces operating on it.
Reinforced masonry may be used as a primary structural system and
can be designed to resist earthquake forces.

## 10.2. Strength of masonry for seismic design

Permissible stresses are laid down in various codes (Building
Seismic Safety Council, 1985; International Conference of Building
Officials, 1985; Standards Association of New Zealand, 1985) and are
related to the level of design forces laid down in each code. Because
masonry varies greatly in quality throughout the world, ultimate
stresses should be based on test results, coupled with a realistic
assessment of the quality of workmanship that can be achieved in
practice.

A useful approximate relationship is to take the permissible
masonry stresses for unreinforced masonry as 0·4 times those for reinforced masonry.

Because test results for masonry prisms are not always available,
Priestley (1985a,b) recommends the following relationship for grouted
concrete masonry

$$f_m' = 0.45\alpha f_{cb}' + 0.675(1 - \alpha)f_g' \tag{10.1}$$

where $f_m'$ is the prism strength of masonry, $f_{cb}'$ is the compressive strength of the concrete masonry unit, $f_g'$ is the compressive strength of the grout and $\alpha$ is the ratio of net to gross area for the masonry unit.

Generally, where the shear stresses for unreinforced masonry are exceeded, the codes require that the whole of the shear is resisted by reinforcement.

### 10.3. Unreinforced masonry

Reinforced masonry is defined by the *Uniform building code* as having an area of reinforcement, both horizontal and vertical, that is greater than 0.0007 times the gross section area, with the sum of both horizontal and vertical reinforcement being greater than 0.002 times the section area.

The seismic capacity for unreinforced masonry is most commonly based on stability and energy considerations rather than stress levels. Neither elastic nor ultimate strength analysis adequately predicts the seismic capacity—both methods produce overconservative results.

Figure 10.1 shows the force–displacement relationship for a masonry wall subjected to static lateral loading. The wall behaves elastically up to point A where the base cracks and the force immediately drops from $F_A$ to $F_B$. Resolving the static forces at the cracked condition

$$F_B = \frac{(P + W)b}{2h} \tag{10.2}$$

The force $F$ reduces to zero at C where, for small rotations

$$Fh = W\left(\frac{b}{2} - \frac{x}{2}\right) + P\left(\frac{b}{2} - x\right)$$

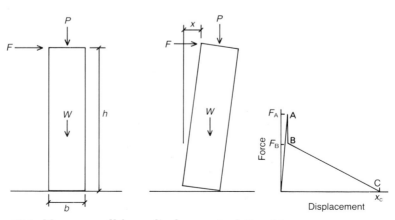

*Fig. 10.1. Masonry wall force–displacement relationship*

from which

$$x = \frac{Wb + Pb - 2Fh}{2P + W} \qquad (10.3)$$

From Fig. 10.1, at point A, the incremental stiffness of the wall becomes negative so that for a steadily applied force $F_A$ collapse will occur unless the force $F_A$ is transferred by an alternative load path to other, stiffer, structural elements. For a ground acceleration pulse this is not necessarily the case, because the pulse which initiated rocking will have to be continued for a sufficient time to reach failure. If the ground acceleration reverses soon after rocking has started, the wall will stabilise again. Under earthquake loading, the displacement may even exceed $x_c$ and return to a stable state if a sufficiently strong reverse pulse occurs.

It has been shown by a number of analytical and practical studies that failure in masonry is closely related to energy. Furthermore it has been demonstrated that the energy requirement to cause in-plane stability failure of masonry walls is so high that failure is normally by shear rather than by instability.

Out-of-plane failure, however, is customarily by instability and Fig. 10.2 shows the simplified response of the masonry wall in Fig. 10.1 to cyclic loading. For simple structures ultimate loading can be assessed approximately by the reserve energy technique (Wesley, Kennedy & Richter, 1980) which replaces the wall by an equivalent elastic structure whose response is shown by the broken line. The energy required to cause failure is approximately equal for the actual and equivalent structures, $x_b$ being much smaller than $x_c$ for practical conditions. The equivalent stiffness can be found from

$$k_e = F_b/x_c \qquad (10.4)$$

and the equivalent natural frequency

$$f_e = \frac{1}{2\pi} \left( \frac{k_e}{W + P} \right)^{1/2} \qquad (10.5)$$

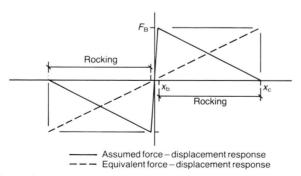

—— Assumed force – displacement response
－－－ Equivalent force – displacement response

*Fig. 10.2. Equivalent structure response*

The maximum displacement can be derived from the appropriate response spectrum, as described in Chapter 3, or other methods that are applicable to elastic structures, the criterion for stability being $x \not> x_c$.

For multi-storey unreinforced masonry structures the problem of assessment becomes more complex and reference should be made to the energy approach described in Priestley (1985a,b).

### 10.4. Infill masonry

The full implications of frame–infill masonry design are complex and rarely taken fully into account in practice. Nevertheless a qualitative understanding of these effects by the building designer is essential in achieving a properly conceived and detailed building.

The first requirement is that walls do not collapse from out-of-plane loading, especially where they might affect means of escape or present a similarly unacceptable hazard. This can be realised by the use of a light, two-way reinforcing grid, bonded to, or mechanically located at, the frame. This is known as basketing, illustrated in Fig. 10.3, and only functions to avoid the final stages of collapse.

Figure 10.4 shows the interaction of the undamaged masonry panel with the frame. The masonry acts as a diagonal compression brace in the direction of the arrow, resulting in a substantial stiffening of the frame and a redistribution of bending moments and shears in the frame.

Figure 10.5 shows the effect of the horizontally sheared panel and accompanying rearrangement of the frame forces. Once the panel has sheared the effect of the diagonal compression zone is lost. Fig. 10.6 shows the situation where the masonry does not occupy the whole of the panel.

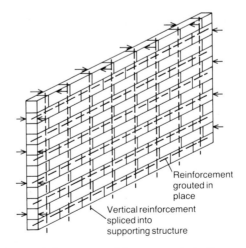

*Fig. 10.3. Basketing reinforcement (the arrows indicate the direction of restraints provided at the frame–wall junction)*

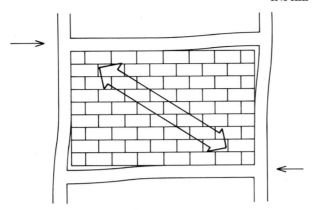

*Fig. 10.4. Interaction between a frame and infill masonry*

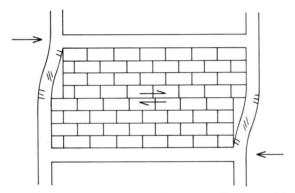

*Fig. 10.5. Interaction between a frame and horizontally sheared infill masonry*

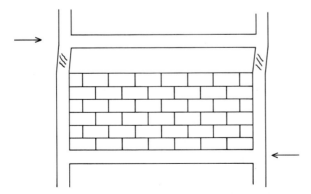

*Fig. 10.6. Interaction between a frame and partial infill masonry*

The redistribution in plan of forces, due to the stiffening effect of infill masonry, is illustrated in Fig. 10.7. Nearly the whole of the lateral force is resisted by the two end frames, being transmitted there by the horizontal stiffness of the floor slab. This situation will continue as long as the masonry panels retain their strength. If masonry at one end only is damaged, high torsional effects will result.

The overall effect of stiffening individual frames with infill masonry is to reduce the natural period of vibration, which (almost invariably) increases the effective lateral force. The local effect of stiffening is to redistribute forces on to the stiffened frames, possibly producing undesirable eccentricity. The consequence may be to increase the forces on a frame many times over.

Contact at the frame–masonry interface modifies the distribution of frame forces. The effective length of a beam or column is shortened so that the ratio of shear to bending force is increased. The reasoning behind this is based on the assumption that the stiffening of one or two members has little effect on the interstorey displacement. The bending moment in the column, $M_{col}$, is given by

$$M_{col} = \frac{6EIx}{l^2} \tag{10.6}$$

The shear force $V_{col}$ is given by

$$V_{col} = \frac{2M}{l} \tag{10.7}$$

so that

$$V_{col} = \frac{12EIx}{l^3} \tag{10.8}$$

where $l$ is the column length and $x$ is the interstorey displacement.

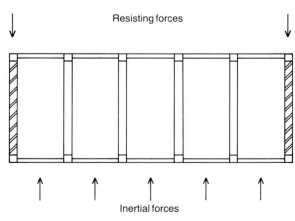

Resisting forces

Inertial forces

*Fig. 10.7. Redistribution of forces on plan due to masonry infill*

From equations (10.6) and (10.8), for example, if a column is braced by masonry over three-quarters of its length, so that only one-quarter of the length is free to displace, the bending moment is increased by 16 times and the shear by 64 times. Clearly the possibility of failure by shear rather than bending is now much greater. In practice numerous cases of shear failures in beams, and more seriously in columns, in this way have been recorded.

A qualitative design approach for the masonry-infilled frame is given by Klingner & Bertero (1977) but is likely to lead to heavy reinforcement, in both the frame and the masonry. The advantage of using this designed infill approach is that full use is made of the additional stiffness provided and of the energy absorbed by reinforced masonry in a major earthquake. An alternative approach is to avoid the uncertainty and complexity involved and to provide a separation between the frame and the masonry infill panel.

A full separation joint between the masonry and frame can be provided at the ends and the top. Out-of-plane failure may then be dealt with either by reinforcing the masonry to act as a vertical canti-

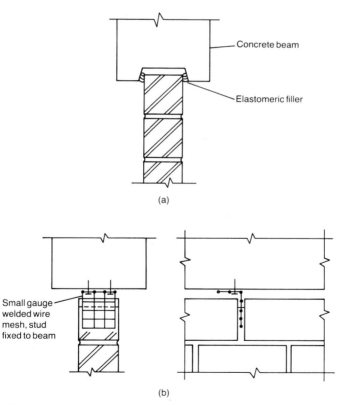

Fig. 10.8. *Lateral restraint details to a 'free' infill panel*

183

lever or by providing basketing reinforcement and mechanically restraining out-of-plane movement at the separation joint. Some suggested details are given in Fig. 10.8.

## 10.5. Structural reinforced masonry

Figures 10.9–10.12 show a number of types of hollow block construction used in reinforced masonry, where reinforcement is grouted in place. The alternative shown in Fig. 10.12 uses a sandwich of masonry-reinforced concrete masonry where the two leaves of masonry act as permanent formwork to the reinforced concrete core.

Patented types of fabricated wire may also be used for horizontal reinforcement where the wire is laid in the horizontal mortar joints.

The design of reinforced masonry follows practice for reinforced concrete walls, using the appropriate lower material stresses. Ductile

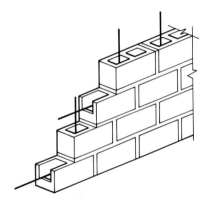

*Fig. 10.9. Alternative types of masonry reinforcement*

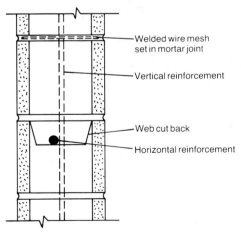

*Fig. 10.10. Alternative types of masonry reinforcement*

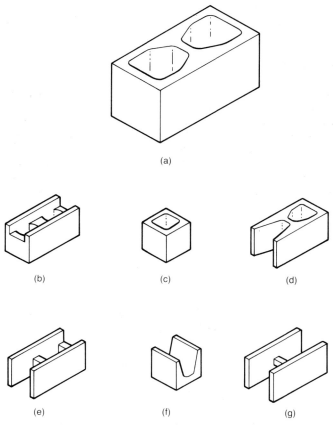

*Fig. 10.11. Standard hollow blocks for reinforced masonry: (a) standard; (b) bond beam; (c) half; (d) open end standard; (e) open end bond beam; (f) lintel; (g) double open end bond beam*

response can be obtained in a similar way to that in reinforced concrete shear walls (see Chapter 7). A comprehensive design guide is given by Englekirk & Hart (1984).

## 10.6. Design example—masonry wall stability
Wall dimensions

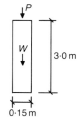

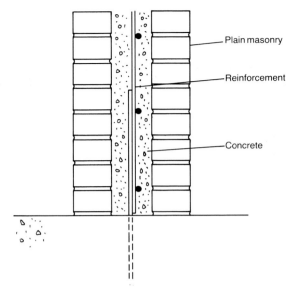

*Fig. 10.12. Solid core reinforced masonry*

For a roof load $P = 15$ kN/m and $W = 7$ kN/m, from equation (10.2)

$$F_b = \frac{(15 + 7) \times 0\cdot15}{2 \times 3} = 0\cdot55 \text{ kN}$$

When $x = x_c$, $F = 0$ and from equation (10.3)

$$x_c = \frac{7 \times 0\cdot15 + 15 \times 0\cdot15}{2 \times 15 + 7} = 0\cdot089 \text{ m}$$

From equation (10.4)

$$k_e = 0\cdot55/0\cdot089 = 6\cdot18 \text{ kN/m}$$

and from equation (10.5)

$$f_e = \frac{1}{2\pi} \left( \frac{6\cdot18 \times 9\cdot81}{15 + 7} \right)^{1/2} = 0\cdot26 \text{ Hz}$$

Using the design spectrum in Fig. 3.6, with a damping coefficient of $0\cdot005$, $x = 3\cdot05$ m for a peak ground acceleration of $1\cdot0g$. Hence the wall will be stable for accelerations up to $0\cdot089/3\cdot05 = 0\cdot029g$.

## 10.7. Bibliography

Amrhein, J. E. (1983). *Reinforced masonry engineering handbook*, 4th edn. Los Angeles: Masonry Institute of America.

Building Seismic Safety Council (1985). *NEHRP recommended provisions for the development of seismic regulations for new buildings*. Washington DC: Building Seismic Safety Council.

Englekirk, R. E. & Hart, G. C. (1984). *Earthquake design of concrete masonry buildings.* Englewood Cliffs: Prentice-Hall.

International Conference of Building Officials (1985). *Uniform building code.*

Klingner, R. E. & Bertero, V. V. (1977). Infilled frames in aseismic construction. *6th World Conf. Earthquake Engineering, New Delhi.*

Priestley, M. J. N. (1985). Seismic behaviour of unreinforced masonry walls. *Bull. N.Z. Soc. Earthquake Engng* **18**, June, No. 2, 191–205.

Standards Association of New Zealand (1985). *Code of practice for masonry design.* Wellington: Standards Association of New Zealand, NZS 4230P.

*Chapter 11*

# Non-structural elements

'The damage suffered by non-structural
components represents a serious threat
to human safety and considerable financial
loss.' Earthquake Engineering Research
Institute report on the Managua earthquake,
Nicaragua, of 23 December 1972

The scope of this chapter covers

(a) the general principles of non-structural element design
(b) static design methods
(c) design practice.

Non-structural masonry is dealt with in Chapter 10

## 11.1. Design objectives

Mechanical equipment, windows, ceilings and cladding may typically represent about 70% of a building's value. The contents can represent many times the value of the building. Failure may involve risk to life, financial loss and the loss of essential post-earthquake services. The preservation of non-structural elements may be equal in importance to maintaining the integrity of the building structure.

In some buildings, e.g. hospitals, it is essential that services are maintained after an earthquake. In such a case much higher standards of design and analysis will be required both for services in the building and external services such as storage tanks, pipelines and substations.

Damage surveys of earthquakes have shown that, in many cases, buildings which have only suffered minor structural damage have been rendered uninhabitable and hazardous to life owing to the failures of mechanical and electrical systems.

## 11.2. General principles

All non-structural items respond to the earthquake motion of the building at the point or points of attachment. Fig. 11.1 illustrates the effect of the building in acting as a filter to the ground motion. Normally the response at any part of the building will be dominated by its own natural frequency and for typical damping values the ground acceleration will be amplified, this amplification increasing with the height above ground.

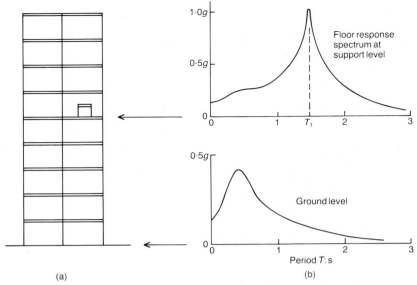

*Fig. 11.1. Secondary structure response to ground motion: (a) multiple degree of freedom structure, first-mode period $T = 1.5$ s; (b) acceleration response spectra*

Analytical methods are dealt with generally in Chapter 3. However, it is possible to proceed without analysis, using equivalent static forces, in a similar procedure to that developed for buildings. Because ground motion is usually amplified by the building, these forces are likely to be considerably higher than those used in the design of the building itself.

The secondary effects of the failure of both mechanical and architectural non-structural elements require careful consideration. Falling masonry or heavy items such as cladding and windows can not only present a hazard to life but may also damage critical mechanical or electrical components. Damaged piping systems, boilers and tanks may explode, or release steam, inflammable, toxic or noxious fluids. Damaged boilers may release materials at a sufficiently high temperature to cause fires.

Although various codes lay down standards for specific categories of building, the application of good engineering judgement in assessing acceptable risks is essential in any project. The traditional division of responsibility between architect, structural engineer and building services engineers does not help this process, and a high degree of coordination between the different disciplines is needed.

### 11.3. Static design method

The *Uniform building code* (International Conference of Building Officials, 1985) adopts the following simple approach to the design of 'parts or portions of structures, non-structural components and their

Table 11.1. Values of $C_p$ for use with equation (11.1)

| Part or portion of building | Direction of force | $C_p$ |
|---|---|---|
| Exterior bearing and non-bearing walls, interior bearing walls and partitions, interior non-bearing walls and partitions | Normal to plane | 0·3 |
| Cantilever elements | | |
| Parapets | Normal to plane | 0·8 |
| Chimneys or stacks | Any direction | 0·8 |
| Exterior and interior ornamentation and appendages | Any direction | 0·8 |
| When connected to, or part of, or housed within, a building: penthouses, anchorage and supports for chimneys, stacks and tanks, including contents storage racks with upper storage level at more than 8 ft (2·44 m) in height, plus contents all equipment or machinery | Any direction | 0·3 |
| Suspended ceiling framing systems | Any direction | 0·3 |
| Connections for prefabricated structural elements other than walls | Any direction | 0·3 |

anchorage to the main structural system'

$$F_p = ZIC_p W_p \qquad (11.1)$$

where $F_p$ is the lateral force acting at the centre of gravity of the element, $Z$ is the seismic zoning coefficient, $I$ is the importance coefficient, $C_p$ is a coefficient with values given in Table 11.1 and $W_p$ is the weight of the element.

$I$ and $Z$ are the values applicable to the building (see Chapter 6) except that $I = 1·5$ for the anchorage of life safety systems.

The *Universal building code* allows $C_p$ to be reduced to two-thirds of the tabulated values for elements whose only lateral support is at ground level, because there is no amplification in this case. For flexible and flexibly mounted machinery and equipment sufficient dynamic analysis is required to establish whether resonance is occurring, the code value of $F_p$ remaining the minimum permissible. This approach is simple and easy to apply, but makes no specific allowance for the possibility of resonance. An alternative approach is given by the Building Seismic Safety Council (1985) as follows.

For architectural components

$$F_p = A_v C_c P W_p \qquad (11.2)$$

where $F_p$ and $W_p$ are the same as in equation (11.1), $C_c$ is a coefficient given in Table 11.2, $A_v$ is the peak ground acceleration (fraction of $g$) and $P$ is the performance criterion factor which varys between 0·5 and 1·5 according to the consequences of failure.

*Table 11.2. Seismic coefficient $C_c$ for architectural systems and components*

| Architectural component | $C_c$ |
|---|---|
| *Appendages* | |
| Exterior non-load-bearing walls | 0·9 |
| Wall attachments | 3·0 |
| Veneers | 3·0 |
| Roofing units | 0·6 |
| Containers and miscellaneous free-standing components | 1·5 |
| | |
| *Partitions* | |
| Stairs and shafts | 1·5 |
| Elevators and shafts | 1·5 |
| Vertical shafts | 0·9 |
| Horizontal exits including ceilings | 0·9 |
| Public corridors | 0·9 |
| Private corridors | 0·6 |
| Full height area separation partitions | 0·9 |
| Other partitions | 0·6 |
| | |
| Structural fireproofing | 0·9 |
| | |
| *Ceilings* | |
| Fire-rated membrane | 0·9 |
| Non-fire-rated membrane | 0·6 |
| | |
| Architectural equipment—ceiling, wall or floor mounted | 0·9 |

For mechanical and electrical systems

$$F_p = A_v \, C_c \, P a_c \, a_x W_c \tag{11.3}$$

where $C_c$ is a coefficient given in Table 11.3 and $a_c$ is the attachment amplification and

$$a_x = 1.0 + \frac{h_x}{h_n} \tag{11.4}$$

where $h_x$ is the height above the base to level $x$ and $h_n$ is the height above the base to roof level.

The attachment amplification factor $a_c$ is defined according to the value of $T_c$ and $T$ where $T_c$ is the fundamental period of the component and $T$ is the fundamental period of the building. When $T/T_c$ lies outside the range 0·6–1·4, or the component is rigidly attached to the building, $a_c = 1.0$. If $T_c/T$ lies within the range 0·6–1·4

$$a_c = \frac{1}{\{[1 - (T/T_c)^2]^2 + (0.04\,T/T_c)^2\}^{0.5}} \tag{11.5}$$

$(a_c \not< 2·0)$.

192

*Table 11.3. Seismic coefficient $C_c$ for mechanical and electrical components\**

| Mechanical/electrical component | $C_c$ |
|---|---|
| Emergency electrical systems | 2·0 |
| Fire and smoke detection systems | 2·0 |
| Fire suppression systems | 2·0 |
| Life safety system components | 2·0 |
| Boilers, furnaces, incinerators, water heaters and other equipment using combustible or high temperature energy sources, chimneys, flues, smokestacks and vents | 2·0 |
| Communication systems | 2·0 |
| Electrical bus ducts and primary cable systems | 2·0 |
| Electrical motor control centres, motor control devices, switchgears, transformers and unit substations | 2·0 |
| Reciprocating or rotating equipment | 2·0 |
| Tanks, heat exchangers and pressure vessels | 2·0 |
| Utility and service interfaces | 2·0 |
| Manufacturing and processing machinery | 0·67 |
| Lighting fixtures | 0·67 |
| Ducts and piping distribution systems | |
|     resiliently supported | 2·0 |
|     rigidly supported | 0·67 |
| Electrical panel boards and dimmers | 0·67 |
| Material conveyor systems | 0·67 |

\* $C_c$ values listed are for horizontal forces. $C_c$ values for vertical forces are taken as one-third of the horizontal values. Hanging or swinging types of fixture have a $C_c$ value of 1·5 and have a safety cable attached to the structure and the fixture at each support point, which is capable of supporting 4·0 times the vertical load.

Equation (11.5) assumes a component damping of 2% critical and gives a peak value at resonance $(T = T_c)$ of 25·0. It is unlikely that such a value would be accepted in practice and alternative types of support would be used.

## 11.4. Planning and detailing

The first and most important requirement for all non-structural components is that they are positively anchored to the building structure. Many important items fail in earthquakes by sliding, toppling or rocking and moving across the floor. Prevention of this type of loss is almost invariably simple and inexpensive. Floor anchorages and angle ties linking the tops of tall items to the structure are usually all that is needed.

A consideration of equation (11.4) makes it clear that mechanical equipment is better located at the lower levels of the building where lower forces are encountered. The desirability of separating the natural frequency of flexible or flexibly supported components from that of the building is clear from equation (11.5). This should be done in a conservative manner as theoretical predictions of the natural frequency may vary substantially from actual values.

Many non-structural components are connected at different levels, so that in addition to resisting applied accelerations they must accommodate differential displacements without failure. Windows, for example, frequently need to accommodate interstorey displacements without failure or fracturing glass. A provision for movement of glass within frames to accommodate racking distortions of 0·5% is advisable and the connection of frames to the structure should provide for yielding of a similar amount. Rigid cladding elements such as precast concrete require similar treatment where they are connected at more than one level. Pipework and ducts traversing movement joints in the building also need to accommodate movement without failure.

Where the structural design is based on equivalent static lateral forces it is important to remember that actual displacements may be ductile and larger than those calculated for the elastic system by a factor typically between 2 and 4.

In order to give some guidance on the principal areas of concern for non-structural items, statistical summaries of damage from major earthquakes have been assembled by Yanev (1981) and Yanev, Swan & Smith (1984), and a brief abstract of significant earthquake-sensitive components is given in Table 11.4.

## 11.5. Isolation for plant

Although experience with spring-supported plant in major earthquakes shows that it may be vulnerable to large displacements and damage, properly designed isolation systems will provide a valuable form of protection. The principles used in design are those described in Chapter 4, taking into account that the base motion is that of the building at the point or points of support.

*Table 11.4.  Common earthquake failures in buildings*

| Item | Type of damage |
|---|---|
| Pumps and boilers | Movement of unanchored supports |
| Tanks | Support failure |
| Motor generators | Failed isolation supports |
| Control panels | Overturning of tall units |
| Piping | Rupture due to excessive movement |
| | Failure at bends |
| Elevators (traction type) | Guide rails broken |
| | Counterweights misaligned |
| | Car misaligned |
| Parapets | Toppling |
| Concrete, stone cladding | Separation and falling |
| Windows | Glass breaking |
| | Frames detaching |
| Storage racks | Toppling and/or contents falling |
| False ceilings | Racking, panels falling |
| Suspended light fittings | Excessive movement causing damage or falling |

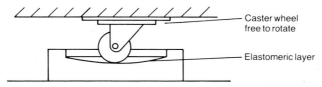

*Fig. 11.2. Caster cup isolating support for equipment (Quantech Systems, California)*

A simple method of sliding isolation, which is suitable for heavy, stable items, is to support them on casters. These may be subjected to substantial displacements relative to the floor. An improvement on this is to use the caster cup which is shown in Fig. 11.2. This is also a simple method, which limits displacement, and by virtue of an elastomeric layer over the caster contact area provides additional damping between the caster and the support.

## 11.6.  Bibliography

Anand, Y. N. (ed.) (1985). *Seismic experience data—nuclear and other plants*. New York: American Society of Civil Engineers.

Building Seismic Safety Council (1985). *NEHRP recommended provisions for the development of seismic regulations for new buildings*. Washington DC: Building Seismic Safety Council.

McGavin, G. L. (1981). *Earthquake protection of essential building equipment*. New York: Wiley.

# Chapter 12

# Non-building structures: a guide

'In my father's house are many mansions.'
*The Gospel of St John*, XIV.2

The scope of this chapter covers

(*a*) bridges
(*b*) tanks
(*c*) chimneys and towers
(*d*) buried pipelines
(*e*) low-rise housing.

## 12.1. Design objectives

Building design requirements are often, erroneously, applied to other types of structure, where both the response and the design objectives may be different from those of a building.

The intention of this chapter is to provide a guide to designers on the types of structure that may be required as ancillary to a building. Additionally further consideration is given to low-rise structures because their design is based less on analysis and more on consider- ations of practical detailing. No consideration is given here to highly specialised areas such as offshore platforms, power lines, nuclear and other high risk industrial structures, dams, embankments or sub- merged structures.

## 12.2. Bridges

In the 1971 San Fernando earthquake, California, damage to highway bridges amounted to US $6 500 000, concentrated within the area of highest seismicity. The prime cause identified in post- earthquake studies was deficient detailing, especially at connections.

The seismic analysis of bridges differs from that for buildings, although the basic principles are similar. The differences are that it will be necessary to model movement joints, and for long spans allow- ance will be needed for differential ground movements at the sup- ports. The same fundamental use of ductility is made in bridges as in buildings, although greater attention is paid to the subsequent repair- ing of the ductile sections—these should be located in visible and accessible places.

Particular attention is paid to holding-down bolts as experience has shown that they are liable to failure in a brittle manner from shear or, less commonly, tension. Accordingly it is normally required that they are designed to remain elastic at ultimate load conditions, or that they can behave in a ductile manner by yielding in bending rather than shearing. Holding-down bolts should be accessible for repair.

One of the prime sources of earthquake failure in bridges has been movement at joints, to the extent that in many cases bridge decks have fallen from their supports. Because movement joints are essential in all but the smallest bridges, it is necessary to provide movement restrainers which will limit excessive movement in an earthquake at both abutments and piers. Movement in ductile structures will be many times larger than that calculated for elastic design levels. A reasonable allowance for movement is

$$\text{displacement} = (\text{elastic displacement}) \times \mu \times 1\cdot5$$

where $\mu$ is the ductility factor.

Movement restrainers may be flexible steel cables or yielding steel links between the bridge deck and supporting structure. Alternatively reinforced concrete keys may be used. Where the possibility of impact exists, usually only in a major earthquake, elastomeric buffers should be used. Provision should also be made for uplift forces developed at joints, based on the capacity design principles described in Chapter 3. Elastomeric bearings are often required to transmit high seismic shears and should be provided with sufficient anchor bolts for this. Fig. 12.1 shows a typical bridge movement joint detail for seismic design.

Because bridge structures supported on abutment walls are laterally very rigid, it may be necessary to undertake a soil–structure interaction analysis. Long span bridges will have low frequencies of vibration, at levels where normal design spectra are unreliable, and a more detailed consideration of ground motion may be necessary.

Bridges are particularly suitable structures for isolation and damping, and many bridges in New Zealand have been designed and

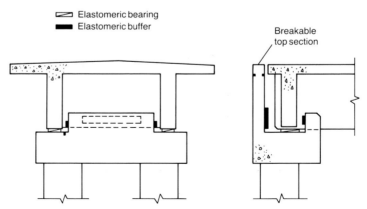

Fig. 12.1. Bridge abutment detail (after Lanigan, Preston Fisher & Stockwell (1980))

constructed on this basis. Chapter 4 deals with this subject in more detail.

## 12.3. Tanks

Liquids stored in tanks will slosh from side to side in earthquakes and exert hydrodynamic pressures on the floor and sides of the tank. The resulting overturning forces lead to the types of failure illustrated in Fig. 12.2. Compression in the walls of a tank due to the overturning forces leads both to the buckling failure in flexible tanks

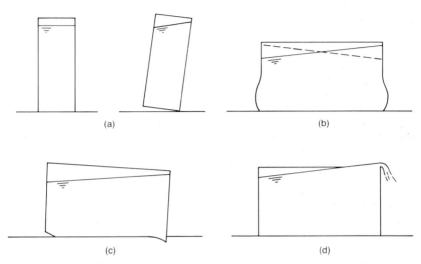

*Fig. 12.2. Tank failure modes: (a) tall tanks overturning; (b) flexible bulk tanks—buckling of the wall ('elephant's foot' failure); (c) flexible bulk tank lifting and ground penetration (both types of movement can lead to tearing of the tank floor); (d) flexible and rigid bulk tanks—overflow*

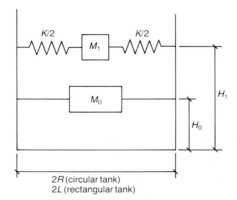

*Fig. 12.3. Tank model (H is the height of stored liquid)*

('elephant's foot') and to tearing at the wall–floor junction. Both these types of failure apply to flexible tanks only.

Bulk tanks are frequently constructed on a compacted soil, or gravel base with a flexible floor of steel plate. Compressive forces due to overturning in the tank wall cause local soil failure where the wall meets the ground, leading to tearing of the floor. This can be avoided by the use of a perimeter foundation to which the tank is bolted. The mass of the foundation is required to be sufficient to avoid uplift. This is a relatively inexpensive provision that is often omitted.

Rigid tanks plus their contents can be modelled on the method proposed by Housner (1963), as illustrated in Fig. 12.3. Values for a circular tank are given by

$$M_0 = M \frac{\tanh(1 \cdot 7R/H)}{1 \cdot 7R/H} \tag{12.1}$$

$$M_1 = M \frac{0 \cdot 71 \tanh(1 \cdot 8H/R)}{1 \cdot 8H/R} \tag{12.2}$$

$$H_0 = 0 \cdot 38H \left[ 1 + \alpha \left( \frac{M}{M_0} - 1 \right) \right] \tag{12.3}$$

$$H_1 = H \left\{ 1 - 0 \cdot 21 \frac{M}{M_1} \left( \frac{R}{H} \right)^2 + 0 \cdot 55\beta \frac{R}{H} \left[ 0 \cdot 15 \left( \frac{RM}{HM_1} \right)^2 - 1 \right]^{1/2} \right\}$$

$$\tag{12.4}$$

$$K = \frac{4 \cdot 75 g M_1^2 H}{MR^2} \tag{12.5}$$

For a rectangular tank the equations are

$$M_0 = M \frac{\tanh(1 \cdot 7L/H)}{1 \cdot 7L/H} \tag{12.6}$$

$$M_1 = M \frac{0 \cdot 83 \tanh(1 \cdot 6H/L)}{1 \cdot 6H/L} \tag{12.7}$$

$$H_0 = 0 \cdot 38H \left[ 1 + \alpha \left( \frac{M}{M_0} - 1 \right) \right] \tag{12.8}$$

$$H_1 = H \left\{ 1 - 0 \cdot 33 \frac{M}{M_1} \left( \frac{L}{H} \right)^2 + 0 \cdot 63\beta \frac{L}{H} \left[ 0 \cdot 28 \left( \frac{ML}{M_1 H} \right)^2 - 1 \right]^{1/2} \right\}$$

$$\tag{12.9}$$

$$K = \frac{3 g M_1^2 H}{ML^2} \tag{12.10}$$

For both conditions, $\alpha = 1\cdot33$ and $\beta = 2\cdot0$ if the moment on the tank bottom is included in the calculation. If only the pressures on the walls are being calculated, $\alpha = 0$ and $\beta = 1$. The amplitude of the waves generated can be taken as the horizontal displacement of $M_1$, $x$, times a factor

$$\eta = \frac{0\cdot69KR/M_1g}{1 - 0\cdot92(x/R)(KR/M_1g)^2} \tag{12.11}$$

for circular tanks and

$$\eta = \frac{0\cdot84KL/M_1g}{1 - (x/L)(KL/M_1g)^2} \tag{12.12}$$

for rectangular tanks.

Damping is effectively zero for practical cases.

The assumptions made for rigid tanks are not valid for flexible tanks and reference should be made to Haroun & Housner (1981) for a design basis. A design code for flexible (welded steel) tanks is given by the American Pipe Institute (1980). Shapes of tank not covered by the standard cases will generally require finite element modelling of the structure and fluid.

Connections of pipework to tanks represent a hazard where possible movement may occur, especially when cast iron valves or fittings are used, and provision for such movement should be made.

## 12.4. Chimney stacks and towers

This section deals with self-supporting chimney stacks and similar tower structures. The design of cable-stayed and frame-supported chimney stacks is generally dominated by wind forces.

Because self-supported stacks have limited redundancy with only a single load path to ground it is not practical to design for ductile behaviour. They should be required to respond elastically to the maximum design earthquake.

An equivalent lateral force design method is provided by ACI 307-69 (American Concrete Institute, 1969). Alternatively response spectrum analysis may be used, in which case at least three modes should be incorporated to allow for the effect of higher modes at the tip of the structure. Several studies have shown that the effect of the elasticity of the ground below the foundation should be included in the model as this has a considerable effect in reducing the modal frequencies. A typical model for analysis is shown in Fig. 12.4, and the calculation of suitable spring and damper values is dealt with in Chapter 9.

ACI 307-69 gives an approximate formula for the natural period of vibration $T$ as

$$T = \frac{0\cdot49H^2}{3D_b - D_t}\left(\frac{m_1}{Em}\right)^{1/2} \tag{12.13}$$

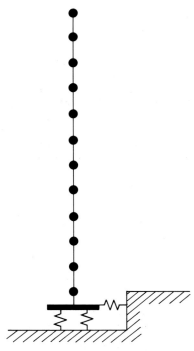

*Fig. 12.4. Structural model for a chimney stack*

where $H$ is the height in metres, $D_b$ is the external diameter at the base in metres, $D_t$ is the external diameter at the top in metres, $E$ is the modulus of elasticity in newtons per millimetre squared, $m_1$ is the total mass of the chimney, $m$ is the total mass of the chimney structure and $T$ is the natural period of vibration in seconds.

A similar formula for chimneys of any material, quoted by Rinne (1970), is

$$T = 5 \cdot 61 \, \frac{wH^4}{EIg} \qquad (12.14)$$

where $H$ and $E$ are defined in equation (12.13), $w$ is the mass per unit of height in kilograms per metre, $D$ is the mean diameter in metres and $I$ is the section's moment of inertia in metres to the fourth power.

Another approximate formula is given by Housner & Keightley (1963) for tapered stacks. A useful discussion of the available approximate formulae is given by Rinne (1970). However, none of these formulae take the effect of soil–structure interaction into account, so they should only be used for preliminary analysis of structures founded on firm ground or rock.

Damping is generally taken, for elastic design levels, as 0·05 for concrete and 0·02 for steel.

## 12.5.  Buried pipelines

The response of buried pipelines to permanent soil movement, such as that due to settlement, liquefaction or fault displacement, may be large and can only be dealt with by mechanical provisions for the movement. Where the integrity of the surrounding soil is maintained pipelines will comply with the soil movement except for axial slippage when the friction grip of the soil on the pipeline surface is exceeded. The results of this are that the predominant effect will be axial strain in the free field and bending at connections.

Generally, pipelines in seismic zones should be flexibly jointed and the prime concern will be in avoiding axial overstressing between joints, avoiding separation or impact at joints and limiting bending stresses at or near points of restraint.

In the free field axial strain $\varepsilon_{amax}$ is given by

$$\varepsilon_{amax} = V_{max}/C \qquad (12.15)$$

where $V_{max}$ is the maximum ground velocity and $C$ is the velocity of wave propagation, and bending strain is given by

$$\varepsilon_{bmax} = \frac{a_{max}}{C^2} \qquad (12.16)$$

where $a_{max}$ is the maximum ground acceleration.

The axial stress due to differential end displacement

$$f_a = \left(\frac{2EF\,\Delta x}{A}\right)^{1/2} \qquad (12.17)$$

where $F$ is the frictional force per unit length, $E$ is Young's modulus, $A$ is the section's area, $\Delta x$ is the differential movement on the long axis and $F = 2\pi d\gamma H f_f$ with $d$ the pipe diameter, $\gamma$ the soil's density, $H$ the depth below the surface and $f_f$ the coefficient of friction, from pipe to soil.

Where a pipe enters a fixed point, the bending stress

$$f_b = \frac{kd\,\Delta x}{4\lambda^2 I} \qquad (12.18)$$

where $k = dk_s$ is the spring constant of the soil perpendicular to the pipeline, $k_s$ is the coefficient of subgrade reaction of soil for a beam of width $d$, along the axial direction and

$$\lambda = \left(\frac{k}{4EI}\right)^{1/4} \qquad (12.19)$$

$I$ is the second moment of area of the pipeline section and $\Delta x$ is the differential movement in the transverse direction.

Guidance on the practical calculation of stresses and movements in jointed pipelines is given by Rascon & Munoz (1984).

## 12.6.  Low-rise housing

For economic reasons the building of low-rise housing is, almost universally, carried out without engineering input. Moreover the analytical approaches that are applicable to larger structures are generally inappropriate.

Regardless of the type of construction there are simple rules that are applicable to the building layout, including reinforced or unre-inforced masonry, quincha (similar to adobe but reinforced with poles), adobe or timber.

(a) The roof structure should be as light as practicable.

(b) The tops of all walls should be tied in a horizontal direction. Normally this is achieved by securely attaching the roof to the walls, which may also be required to achieve sufficient tying-down forces for wind uplift.

(c) On plan the layout of walls should be as symmetric as possible about each axis. Simple box shapes are preferable to complex shapes.

(d) Openings in exterior walls should be at least 500 mm from the corners.

(e) Interior doorways should be at least two wall thicknesses away from the end of the wall.

(f) Openings in walls should be at least 500 mm apart.

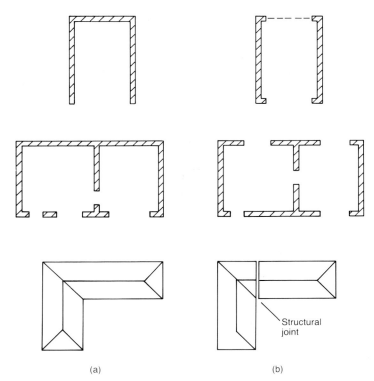

(a)                                        (b)

Fig. 12.5. Design guide for low-rise dwellings: (a) non-symmetric plan; (b) modified plan for near symmetry

(g) The total width of openings in an external wall should not exceed one-third of the length of the wall.

(h) Unreinforced parapet walls or other projections such as chimneys should not be built.

These rules are illustrated in Figs 12.5 and 12.6.

Although unreinforced masonry should generally not be acceptable for earthquake areas, economic considerations result in wide usage. Almost any reinforcement is better than none. Vertical reinforcement is particularly desirable at corners and some applications for hollow and solid masonry are shown in Fig. 12.7. A reinforced concrete capping beam is a further improvement.

Although timber is in many ways an ideal material for earthquake construction, being light, strong and capable of absorbing energy, a surprising number of timber-framed houses were badly damaged in the 1971 San Fernando earthquake, Los Angeles. This was generally due to poor detailing. It is essential for timber construction that racking failures are avoided by the inclusion of sufficient diagonal

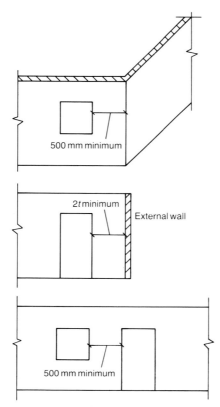

Fig. 12.6. Design guide for openings in masonry house walls (after Daldy (1972)

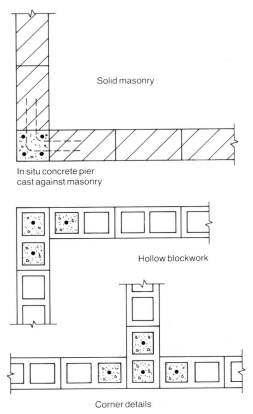

Solid masonry

In situ concrete pier
cast against masonry

Hollow blockwork

Corner details

*Fig. 12.7. House masonry details*

bracing and that tying-down bolts at foundation level are adequate.

Many low-rise houses have the floor above ground level and supported on piers and isolated pad footings. This is a poor seismic
concept and requires properly engineered design for lateral forces.

## 12.7. Bibliography

American Association of State Highway and Transportation Officials,
Subcommittee on Bridges and Structures (1981). *Interim specifi-
cations: bridges.* Washington DC: AASHTO.

American Concrete Institute (1969). Specification for the design and
construction of reinforced concrete chimneys, ACI 307-69. *Am.
Concr. Inst.*

American Pipe Institute (1980). *Welded steel tanks for oil storage.* API
Standard 650, 7th edn. Washington DC: API.

Applied Technology Council (1981). *Seismic design guidelines for
highway bridges.* Report No. ATC-6. Redwood City: ATC.

New Zealand Society for Earthquake Engineering (1980). Seismic
design of bridges. *Bull. N.Z. Soc. Earthquake Engng* **13**, Sept. No. 3.

# References

Adeli, H., Gere, J. M. & Weaver, W. (1978). Algorithms for non-linear structural dynamics. *J. Struct. Div. Am. Soc. Civ. Engrs* **104**, Feb., 263–280.

American Concrete Institute (1969). *Standard specification for the design and construction of reinforced concrete chimneys.* ACI 307–69. Detroit: ACI.

American Concrete Institute (1983). *Standard building code requirements for reinforced concrete.* ACI 318–83. Detroit: ACI.

American Concrete Institute–American Society of Civil Engineers, Committee 352 (1985). *Recommendations for design of beam–column joints in monolithic reinforced concrete structures.* ACI 352R-85. Detroit: American Concrete Institute.

American Institute for Steel Construction (1984). *Engineering for steel construction.* Chicago: AISC.

American Pipe Institute (1980). *Standard for welded steel tanks for oil storage.* API 650, 7th edn. Washington DC: API.

American Society of Civil Engineers (1980). Structural analysis and design of nuclear plant facilities. *Man. Rep. Engng Practice*, No. 58.

Anand, Y. N. (ed.) (1985). *Seismic experience data—nuclear and other plants.* New York: American Society of Civil Engineers.

Applied Technology Council (1981). *Seismic design guidelines for highway bridges.* ATC 6. Redwood City: ATC.

Applied Technology Council (1982). *Tentative provisions for the development of seismic regulations for buildings, 1978.* ATC 3-06. Redwood City: ATC.

Arnold, C. (1984). Soft first stories: truths and myths. *Proc. 8th World Conf. Earthquake Engineering, San Francisco* **5**, 943–949. Englewood Cliffs: Prentice Hall.

Bathe, K.-J. & Wilson, E. L. (1976). *Numerical methods in finite element analysis.* Englewood Cliffs: Prentice-Hall.

Blakely, R. W. G., Charleson, A. W., Hitchcock, H. C., Megget, L. M., Priestley, M. J. N., Sharpe, R. D. & Skinner, R. I. (1979). Recommendations for the design and construction of base isolated structures. *Bull. N.Z. Soc. Earthquake Engng* **12**, June, No. 2, 136–157.

Blume, J. A., Newmark, N. M. & Corning, L. H. (1961). *Design of multistory reinforced concrete buildings for earthquake motions.* Chicago: Portland Cement Association.

British Standard BS6177: 1982. *Guide to selection and use of elastomeric bearings for vibration isolation of buildings.* London: BSI.

Broms, B. B. (1965). Design of laterally loaded piles. *J. Soil Mech. Fdns Div. Am. Soc. Civ. Engrs* **91**, May, No. 3, 79–99.

Building Seismic Safety Council (1985). *NEHRP recommended provisions for the development of seismic regulations for new buildings.* Washington DC: BSSC.

Butterworth, J. W. & Spring, K. C. F. (1985). Seismic design of steel structures. *Bull. N.Z. Soc. Earthquake Engng* **18**, Dec., No. 4, 344–350.

## REFERENCES

Cardenas, A. E., Hanson, J. M., Corley, W. G. & Hognestad, E. (1973). Design provisions for shear walls. *A.C.I. J.* **70**, Mar., No. 3, 221–230.

Comité Euro-International du Breton (1985). *Model code for seismic design of concrete structures.* EPFL-CP88-CH-1015. Lausanne: CEB.

Chopra, A. K., Clough, D. P. & Clough, R. W. (1973). Earthquake resistance of buildings with a 'soft' first story. *Earthquake Engng Struct. Dynam.* **1**, No. 4, 347–355.

Clough, R. W. & Penzien, J. (1975). *Dynamics of structures.* New York: McGraw Hill.

Daldy, A. F. (1972). *Small buildings in earthquake areas.* Watford: Building Research Establishment.

Degenkolb, H. J. (1978). Steel connections for earthquake resistance. *Spring Conv., Pittsburgh.* Preprint 3177. New York: American Society of Civil Engineers.

Derham, C. J. (1986). Non-linear natural rubber bearings for seismic isolation. *Proc. Sem. Workshop Base Isolation and Passive Energy Dissipation, San Francisco.* Redwood City: Applied Technology Council.

Douty, R. T. & McGuire, W. (1965). High strength bolted moment connections. *J. Struct. Div. Am. Soc. Civ. Engrs* **91**, Apr., No. 2, 101–128.

Earthquake Engineering Research Laboratory (1980). *Earthquake strong motion records.* Report No. 80-01. Pasadena: EERL.

Englekirk, R. E. & Hart, G. C. (1984). *Earthquake design of concrete masonry buildings* **1**, **2**. Englewood Cliffs: Prentice-Hall.

Fédération Internationale de la Précontrainte (1977). *Recommendations for the design of aseismic prestressed concrete structures.* Slough: Cement & Concrete Association.

Fédération Internationale de la Précontrainte (1978). Chairman's report. *Proc. 8th FIP Congr., London* Part 2, 45–50. FIP.

Ferrito, J. M. (1984). Economics for seismic design for new buildings. *J. Struct. Div. Am. Soc. Civ. Engrs* **110**, Dec., No. 12, 2925–2938.

Fintel, M. & Khan, F. R. (1969). Shock absorbing soft story concept for multi-story earthquake structures. *A.C.I. J.*, May, 381–390.

Gosstroi, USSR (1969). *Construction in earthquake regions.* (English trans.) Watford: Building Research Establishment.

Hadjian, A. H. (1982). A re-evaluation of equivalent linear models for simple yielding systems. *Earthquake Engng Struct. Dynam.* **10**, 759–767.

Hardin, B. O. & Richart, F. E. (1963). Elastic wave velocities in granular soils. *J. Soil Mech. Fdns Div. Am. Soc. Civ. Engrs* **97**, No. 9, 1249–1273.

Hardin, B. O. & Drnevich, V. P. (1972a). Shear modulus and damping in soils: measurement and parameter effects. *J. Soil Mech. Fdns Div. Am. Soc. Civ. Engrs.* **98**, June, No. 6, 603–624.

Hardin, B. O. & Drnevich, V. P. (1972b). Shear modulus and damping in soils: design equations and curves. *J. Soil Mech. Fdns Div. Am. Soc. Civ. Engrs* **98**, July, No. 7, 667–692.

Haroun, M. A. & Housner, G. W. (1981). Seismic design of liquid storage tanks. *J. Tech. Counc. Am. Soc. Civ. Engrs* **107**, 191–207.

Hjelmstad, K. D. & Popov, E. P. (1984). Characteristics of eccentrically braced frames. *J. Struct. Div. Am. Soc. Civ. Engrs* **110**, Feb., No. 2, 340–353.

Houbolt, J. C. (1950). A recurrence matrix solution for the dynamic response of elastic aircraft. *J. Aero. Sci.* **17**, 540–550.

Housner, G. W. (1963). Dynamic behaviour of water tanks. *Bull. Seismol. Soc. America* **53**, No. 2, 381–387.

Housner, G. W. & Jennings, P. C. (1982). *Earthquake design criteria.* Berkeley: Earthquake Engineering Research Institute.

208

Housner, G. W. & Keightley, W. O. (1963). Vibrations of linearly tapered beams. *Trans. Am. Soc. Civ. Engrs* **128**, Part 1, 1020–1048.

Hutchinson, G. L. & Chandler, A. M. (1986). Parametric earthquake response of torsionally coupled buildings and comparison with earthquake codes. *Proc. 8th Eur. Conf. Earthquake Engng, Lisbon* **3**, 25–32.

International Conference of Building Officials (1982). *Uniform building code*, Whittier, CA.

International Conference of Building Officials (1985). *Uniform building code*. Whittier, CA.

Iwan, W. D. (1980). Estimating inelastic response spectra from elastic spectra. *Earthquake Engng & Struct. Dynam.* **8**, No. 4, 375–388.

Kanai, K. (1967). Semi empirical formula for seismic characterisation of the ground. *Bull. Earthquake Res. Inst. Univ., Tokyo* **35**, June.

Karamanchandi, A. K. & Reed, J. W. (1986). Validity of approximate analysis techniques for base isolated structures with non-linear isolation elements. *Proc. 3rd US Conf. Earthquake Engng, Charleston* **3**, 1982–1992. Berkeley, CA: Earthquake Engineering Research Institute.

Kasai, K. & Popov, E. P. (1984). On seismic design of eccentrically braced frames. *Proc. 8th World Conf. Earthquake Engng, San Francisco* **5**, 387–394. Englewood Cliffs: Prentice Hall.

Klingner, R. E. & Bertero, V. V. (1977). Infilled frames in aseismic construction. *6th World Conf. Earthquake Engng, New Delhi*. Englewood Cliffs: Prentice Hall.

Lanigan, A. G., Preston, R. L., Fisher, R. W. & Stockwell, M. J. (1980). Seismic design of bridges: section 8, structural and non-structural details. *Bull. N.Z. Soc. Earthquake Engng* **13**, Sept., No. 3, 274–279.

Leslie, S. K. & Biggs, J. M. (1972). Earthquake code evolution and the effect of seismic design on the cost of buildings. *Massachusetts Institute of Technology Structures Publ. 341*. Cambridge: MIT.

Martin, L. D. & Korkosz, W. J. (1982). *Connections for precast concrete buildings*. Prestressed Concrete Institute, Chicago, Technical Report 2.

Mayes, R. L., Jones, L. R., Kelly, T. E. & Button, M. R. (1984). Design guidelines for base isolated buildings with energy dissipators. *Earthquake Spect.* **1**, Nov., No. 1, 41–74.

McKay, G. R. (1985). Seismic design of steel structures. *Bull. N.Z. Soc. Earthquake Engng* **18**, Dec., No. 4, 400–405.

Newland, D. E. (1975). *Random vibrations and spectral analysis*. London: Longman.

Newmark, N. M. (1959). A method of computation for structural dynamics. *J. Engng Mech. Div. Am. Soc. Civ. Engrs* **85**, 67–94.

Newmark, N. M. & Hall, W. J. (1982). *Earthquake spectra and design*. Berkeley: Earthquake Engineering Research Institute.

Newmark, N. M. & Rosenblueth, E. (1971). *Fundamentals of earthquake engineering*. Englewood Cliffs: Prentice-Hall.

New Zealand Ministry of Works and Development (1973). Retaining wall design notes. *Civil Division Publ. CDP 702/C*. Wellington: NZMWD.

New Zealand Ministry of Works and Development (1981). *Seismic design of petrochemical plants*. Wellington: NZMWD.

New Zealand standard code of practice for the design of concrete structures. *NZS 3101:1982, parts 1 and 2*. Wellington: Standards Assoc. of New Zealand.

New Zealand standard code of practice for the design of masonry structures (1985). *NZS 4230P:1985*. Wellington: Standards Assoc. of New Zealand.

Nicholas, C. J. A. (1985). Seismic design of steel structures. *Bull. N.Z. Soc.*

REFERENCES

*Earthquake Engng* **18**, Dec., No. 4, 360–368.

Park, R. & Paulay, T. (1975). *Reinforced concrete structures.* New York: Wiley.

Patton, R. N. (1985). Seismic design of steel structures. *Bull. N.Z. Soc. Earthquake Engng* **18**, Dec., No. 4, 329–336.

Paulay, T. (1983). Deterministic seismic design procedures for reinforced concrete buildings. *Engng Structs* **5**, Jan., 79–86.

Paulay, T. & Goodsir, W. J. (1985). The ductility of structural walls. *Bull. N.Z. Soc. Earthquake Engng* **18**, Sept., No. 3, 250–269.

Priestley, M. J. N. (1985a). Seismic design of masonry structures to the New Zealand Standard NZS 4230P. *Bull. N.Z. Soc. Earthquake Engng* **18**, Mar., No. 1.

Priestley, M. J. N. (1985b). Seismic behaviour of unreinforced masonry walls. *Bull. N.Z. Soc. Earthquake Engng* **18**, June, No. 2, 191–205.

Rascon, O. A. & Munoz, C. J. (1984). Recommendations for seismic analysis of buried pipelines in Mexico City. *Proc. 8th World Conf. Earthquake Engng, San Francisco* **7**, 309–336. Englewood Cliffs: Prentice Hall.

Richards, R. J. & Elms, D. (1985). Seismic behaviour of gravity retaining walls. *J. Geotech. Engng Div. Am. Soc. Civ. Engrs* **105**, Apr., No. 4, 449–464.

Rinne, J. E. (1970). *Design of earthquake resistant structures, towers & chimneys in earthquake engineering (ed: Wiegel, R. L.).* 495–505. Englewood Cliffs: Prentice Hall.

Rutenberg, A. & Heidebrecht, A. C. (1985). Response spectra for torsion, rocking and rigid foundations. *Earthquake Engng Struct. Dynam.* **13**, 543–557.

Salse, E. A. B. & Fintel, M. (1973). Strength, stiffness and ductility properties of slender shear walls. *Proc. 5th World Conf. Earthquake Engng, Rome* **1**, 919–928.

Seed, H. B. (1968). *Characteristics of rock motion during earthquakes.* University of California at Berkeley, Report EERC 63-5.

Seed, H. B. & Idriss, I. M. (1970). *Solid moduli and damping factors for dynamic response analyses.* Report EERC 70-10. University of California at Berkeley: Earthquake Engng Research Center.

Seed, H. B., Idriss, I. M. & Dezfulian, H. (1970). *The relationship between soil conditions and building damage in the Caracas earthquake of July 29th, 1967.* University of California at Berkeley, Report EERC 70-2.

Seed, H. B. & Idriss, I. M. (1971). Simplified procedure for evaluating soil liquefaction potential. *J. Soil Mech. Fdns Div. Am. Soc. Civ. Engrs* **97**, No. 9, 1249–1273.

Seed, H. B., Murarka, R., Lysmer, J. & Idriss, I. M. (1976). Relationships of maximum acceleration, maximum velocity, distance from source, and local site conditions for moderately strong earthquakes. *Bull. Seismol. Soc. Am.* **66**, No. 4, 1323–1342.

Seed, H. B. (1979). Soil liquefaction and cyclic mobility evaluation for level ground during earthquakes. *J. Geotech. Engng Div. Am. Soc. Civ. Engrs* **105**, Feb., No. 2, 201–255.

Seed, H. B. & Idriss, I. M. (1982). *Ground motion and soil liquefaction during earthquakes.* Richmond: Earthquake Engineering Research Institute.

Sherif, M. A. & Fang, Y.-S. (1984). Dynamic earth pressures on rigid walls rotating about the base. *Proc. 8th World Conf. Earthquake Engng, San Francisco* **6**, 993–1000.

Silver, M. L. (1981). Load deformation and strength behaviour of soils under dynamic loading. *Proc. Int. Conf. Recent Advances in Geotech. Earthquake Engng Soil Dynam. University of Missouri-Rolla, Apr.* **3**, 873–890.

Structural Engineers Association of California (1974). *Recommended lateral force requirements.* Sacramento: SEAC.

Structural Engineers Association of California (1985). *Tentative lateral force requirements.* Sacramento: SEAC.

Taylor, P. W. & Williams, R. L. (1979). Foundations for capacity designed structures. *Bull. N.Z. Soc. Earthquake Engng* 12, June, No. 3, 101–113.

Tsai, N. C. (1984). A new method for spectral response analysis. *Proc. 8th World Conf. Earthquake Engng, San Francisco* 4, 171–177. Englewood Cliffs: Prentice Hall.

Tyler, R. G. (1978). Tapered steel energy dissipators for earthquake resistant structures. *Bull. N.Z. Soc. Earthquake Engng* 11, Dec., No. 4, 282–294.

United States Nuclear Regulatory Commission (1976). *Reactor safety study.* Report No. WASH-1400. Washington DC: USNRC.

Vanmarcke, E. H. (1976). *Earthquake risk and engineering decisions* (eds C. Lomnitz and E. Rosenblueth), Ch. 8. Amsterdam: Elsevier.

Walpole, W. R. (1985). Seismic design of structures. *Bull. N.Z. Soc. Earthquake Engng* 18, No. 4, 369–380.

Walpole, W. R. & Butcher, G. W. (1985). Seismic design of structures. *Bull. N.Z. Soc. Earthquake Engng* 18, No. 4, 337–343.

Wesley, D. A., Kennedy, R. P. & Richter, P. J. (1980). Analysis of the seismic collapse capacity of unreinforced masonry wall structures. *Proc. 7th World Conf. Earthquake Engng, Istanbul* 6, 411–418. Tokyo: Int. Assoc. for Earthquake Engng.

Whitman, R. V. & Richart, F. E. (1967). Design procedures for dynamically loaded foundations. *J. Soil Mech. Fdns Div. Am. Soc. Civ. Engrs* 93, 161–191.

Whitman R. V., Biggs, J. M., Brennan, J. E., Cornell, C. A., de Neufville, R. L. & Vanmarcke, E. H. (1975). Seismic design decision analysis. *J. Struct. Div. Am. Soc. Civ. Engrs* 101, May, No. 5.

Whitman, R. V. & Liao, S. (1984). Seismic design of gravity retaining walls. *Proc. 8th World Conf. Earthquake Engng, San Francisco* 3, 533–540.

Wilteveen, J. H., Stark, J. W. B., Bijlaard, F. S. K. & Zoetemeijer, P. (1982). Welded and bolted beam to column connections. *J. Struct. Div. Am. Soc. Civ. Engrs* 108, Feb., No. 2, 433–455.

Wolf, J. P. (1985). *Dynamic soil structure interaction.* Englewood Cliffs: Prentice-Hall.

Wood, J. H. (1973). *Earthquake induced soil pressures on structures.* California Institute of Technology, Pasadena, Report EERL 73-05.

Yanev, P. I. (1981). Effects and implications of past earthquakes on power plants. *Proc. Am. Pwr Conf. Chicago* 43.

Yanev, P. I., Swan, S. W. & Smith, N. P. (1985) A summary of the seismic qualification utilities group (SQUG) program. *Seismic experience data—nuclear and other plants (ed. Anand, Y. N.)* p.p. 1–13. New York: Am. Soc. Civ. Engrs.

# Appendix 1

# Computer programs

There are far too many programs in use for a summary to be practicable. For this reason programs from a single source, the National Information Service on Earthquake Engineering, are given. All the programs listed are available from

> National Information Service on Earthquake Engineering
> Computer Applications
> Earthquake Engineering Research Center
> 379 Davis Hall
> University of California
> Berkeley, CA 94720
> USA

They are available on tape or disk and the listings are given, so that the user may develop or modify them for his own purposes.

*Dynamic structural analysis, linear*

| | |
|---|---|
| COMBAT | SAP4 |
| ETABS | TABS77 |
| SUPER ETABS | TABS80 |

*Dynamic structural analysis, non-linear*

| | | |
|---|---|---|
| ANSR1 | DRAIN2D | NONSAP |
| ANSR1:INEL1 | DRAIN2D:EL9 | SAKE |
| ANSR1:INEL2 | DRAIN2D:EL10 | |
| ANSR1:INEL4 | DRAIN-ETABS | |

*Soil dynamics*

| | |
|---|---|
| CHARSOIL | QUAD4 |
| LUSH2 | SHAKE |
| MASH | |

*Soil liquefaction*

| | |
|---|---|
| APOLLO | LASS2 |
| CUMLIQ | LASS3 |
| GADFLEA | |

*Earthquake risk for a site*

EQRISK

*Seismic evaluation of existing structures*

| | | |
|---|---|---|
| ACE | DAEM | DAMAGE |

COMPUTER PROGRAMS

*Retaining walls*
>BASSIN

*Deconvolution of surface motion to base rock*
>LAYER

*Artificial earthquake generation*
>PSEQGN  SHOCHU  SIMQKE

*Non-linear spectra*
>NONSPEC

*Linear spectra*
>SPECEQ/UQ  SPECTR

*Pile analysis*
>SPASM

Program listings are published in some textbooks, including

(*a*) Bathe, K.-J. & Wilson, E. L. (1976). *Numerical methods in finite element analysis.* Englewood Cliffs: Prentice-Hall.
(*b*) Paz, M. (1980). *Structural dynamics theory and computation.* New York: Van Nostrand Reinhold.

214

# Index